PLASTICS RHEOLOGY

R. S. LENK

PLASTICS RHEOLOGY

Mechanical behaviour of solid
and liquid polymers

WILEY INTERSCIENCE
A DIVISION OF JOHN WILEY & SONS, INC.
NEW YORK

85334 040 4
PLASTICS RHEOLOGY
BY R. S. LENK
PUBLISHED IN THE U.S.A. 1968 BY
WILEY INTERSCIENCE
A DIVISION OF JOHN WILEY AND SONS, INC.
NEW YORK
© 1968
R. S. LENK
LIBRARY OF CONGRESS CATALOG CARD NO. 68-9486

MADE IN GREAT BRITAIN
PRINTED AT THE PITMAN PRESS, BATH

IN MEMORY OF MY UNCLE,

THE LATE

PROF. DR ROBERT LENK,

PIONEER IN RADIOLOGY

Foreword

WHEN at the beginning of the 1964/5 session I was asked to present a series of lectures on Plastics Rheology I was aware of the stream of important papers which had been published at an exponentially increasing rate since the War. I was also aware that there was as yet no general textbook which would provide a unified summary of the state of the science at a level suitable for ready assimilation by students reading for the B.Sc. in Chemical Technology with plastics specialisation or by students preparing for the Associateship examinations of the Plastics Institute by the Physics or Engineering route.

While our students in their final undergraduate year or at post-graduate level could certainly work their way through the major works on rheology, it is scarcely reasonable to expect them to do so, because:

1. Their concern with rheology is centred on *plastics* rheology.

2. The attempt to assimilate existing standard treatises, excellent though they may be, would have encroached unfairly upon the hours of private study available to even a full-time (let alone a part-time) student. The student has to allocate his time with due regard to the claims of other subjects within his chosen discipline, and these may be almost as important as rheology and subjects in which he will certainly have to satisfy a searching examination in his finals.

3. Such textbooks as are available tend to presuppose some familiarity with the terminology and basic concepts of rheology.

4. These textbooks are often so highly mathematical that they are apt to frighten or discourage one who may be essentially a chemist.

5. Existing textbooks often start out on a physicochemical basis the concepts of which are unfamiliar to engineers, even though they may be readily accessible to physicists.

As the lecture series took shape I fell into the habit of issuing 'revision notes' which were intended to remind the students of important features of the lecture. This enabled them to concentrate their attention on the arguments presented and although they took careful notes they were freed from the anxiety of actually missing important steps whilst copying mathematical derivations from the blackboard. They could then go away, secure in the knowledge that they would be able to digest the material presented, the significance of which often does not appear until the hustled concentration of the lecture theatre gives way to the more leisurely reflections of private seclusion.

As time went on, these 'revision notes' became a sizeable loose leaf collection and the continuing absence of a textbook suitable to this level led me to consider the feasibility of filling the gaps, based on two years' lectures, extensive perusal of the literature and several years' industrial rheological research. I approached this ambitious task with the humility befitting one who is well aware of the pioneering work of others who have devoted their entire professional life to rheology. I trust that they will forgive me for borrowing from their writings, that they will feel compensated by suitable acknowledgement and that they may even be so kind as to confer their blessing on an effort made in a cause which is surely dear to them.

I feel that this is also the place where those must be comforted who—in my friend Dr Zamodits' words—'turn green at the gills when the word "tensor" is mentioned.' The scanty treatment which vector and tensor analysis generally receives at school will have to be remedied in the course of periodic revision of sixth-form mathematics syllabuses sooner or later. It is the engineers in particular who suffer from this gap in their basic mathematical education—physicists make it up when they enter university.

I have tried to incorporate an introduction to vectors and tensors in the simplest possible form in Chapters 2 and 3. This is necessary for the rigorous treatment of extruder and calender flow problems. Those readers who are not normally concerned with extruder flow problems may feel inclined to omit these chapters on first reading. Others may wish to accept the final equations which are used for the solution of the specific examples given in Chapter 3 and return to their derivation at a later stage. But it should be stated that, eventually, vector and tensor calculus must be tackled by the rheologist because the separation of the 'theoretician' and the 'practical man' into types which are mutually and unjustly regarded from the one side as ivory tower squatters, and from the other as ignoramuses, *must* be broken down. The creation of a bridge of mutual regard and the establishment of common ground between the two makes it essential that the *basic concepts* of vector and tensor calculus become *lingua franca*. Only thus will the 'theoretician' cease to be viewed with suspicion by the industrial rheologist and only thus will the theoretician begin to feel that his researches are not relegated to a sterile pigeonhole, but understood, appreciated and applied in industry. Eventually this will bring about the desirable state in which the two 'types' will unite in one species.

I acknowledge my debt of gratitude to Mr J. Proctor, Principal Lecturer in Plastics Technology at the Borough

Polytechnic, who has given me encouragement in writing this book. It should be said that I have drawn liberally on the published work of J. F. Carley in Chapter 6, on McKelvey's excellent textbook on Polymer Processing in Chapters 2, 3, and 4 and on the published work of P. Vincent and S. Turner. I am indebted to all these and others who have given me permission to draw upon their work and to the publishers concerned.

I am very conscious that this is still a formative period for plastics rheology and that any textbook written now may well obsolesce fairly rapidly. This indeed is the case with any subject of intense current interest and activity. But the point has been reached when a unified basis can be formulated and the attempt to write a textbook must therefore be made. Whether the scope of this book has been too modest or too ambitious the reader will decide. It could well be that with the present and expected further growth in rheological research, a revision will become necessary within a relatively short space of time. But if there is a span of a few years of usefulness vouchsafed for this book, during which it may serve recruits to plastics rheology at this college and elsewhere, and during which it may stimulate and assist plastics technologists generally (irrespective of their original scientific background), then I will cheerfully bear the prospect of its inherent obsolescence.

Department of Chemistry   R. S. LENK
and Chemical Technology,
Borough Polytechnic,
London, S.E.1.
January 1968

Contents

and the rolling ball spectrometer. Summary on the uses of dynamic spectra (especially electrical and mechanical loss spectra) for structural elucidation and product performance.

Acknowledgements

THE author and publishers wish to express their grateful thanks for permission to reproduce various tables and illustrations as set out below:

R. B. Bird, W. E. Stewart and N. Lightfoot and John Wiley and Sons, Ltd. for: Tables 3.4-1, 3.4-2, 3.4-3, 3.4-4, 3.5-5, 3.4-6, 3.4-7, 3.4-8, 10.2-1, 10.2-2 from *Transport Phenomena* (1960).

P. Clegg and Pergamon Press for: Schematically reproduced photographs of the behaviour of the coloured core of polythene melt in laboratory extrusion from *The Rheology of Elastomers* (1958).

J. F. Carley, *SPE Journal* (Sept. 1963, Dec. 1963) for: Isovels in a rectangular and slit die from 'Rheology and Die Design'. *Modern Plastics Magazine* (Aug. 1956) for: Typical profile dies and calculation of effective slit width; and a typical log/log plot of shear stress vs. shear rate in a low density polythene at various temperatures from 'Problems of Flow in Extrusion Dies.'

S. Turner and *British Plastics* (June 1964) for: Representation of solid model showing a typical stress/strain/time surface from 'Stress/Strain/Time Temperature Relationships and their Evaluation.'

W. Philippoff and F. H. Gaskins and *Journal of Polymer Science*, 1956 (John Wiley and Sons, Ltd.) for: Table for activation energies of flow of molten polythene from 'Activation Energies of Molten Polymers.'

H. J. Scherr and W. E. Palm and *Journal of Applied Polymer Science*, 1963 (John Wiley and Sons, Ltd.) for: Diagrammatic reproduction of an apparatus for the measurement of creep from 'Apparatus for Creep Measurement.'

I. Cheetham and The Institution of the Rubber Industry, Aug. 1965; Messrs. H. W. Wallace and Co., Ltd., Croydon (Technical Literature) for: Diagrams of a rolling ball hysteresis loss spectrometer from 'A Rolling Ball Loss Spectrometer.'

P. Vincent and *Plastics* (Oct. 1961, etc.) for: Various graphs derived from stress/strain plots as detailed where cited; Representation of composite stress systems from 'The Strength of Plastics.'

G. C. Karas and *British Plastics* (Feb. 1964) for: Four graphs (partly after other authors) showing damping as a function of temperature from 'Principles of Dynamic Testing.'

J. R. Van Wazer, J. W. Lyons, K. Y. Kim and R. E. Colwell, and John Wiley and Sons, Ltd. (1963) for: The major part of the diagrammatic representation of various types of shear as in Fig. 1.1 of this book from 'Viscosity and its Measurement.'

A. A. Miller and *Journal of Polymer Science* (June 1963) for: Approximate reproduction of Fig. 4.2 of this book—graphical determination of the reference temperature T_r from 'A Reference point for Molecular Relaxation Processes.'

List of Symbols

A	cross sectional area
A	a constant
A'	derivative of A
A_0	original cross sectional area
B	coefficient of thixotropic breakdown with time
B	a constant
C	a constant
C_v	heat capacity at constant volume
C_0	vacuum capacitance in farads (ohms^{-1} cm^{-2})
D	a dimensionless quantity in extruder equations
D	a state
D	an operator
E	activation energy
E_a	absorbed energy
E_r	recoverable energy
E_τ	activation energy at constant shear stress
$E_{\dot\gamma}$	activation energy at constant shear rate
$F(.\ .\ .)$	Function of $(.\ .\ .)$
F	tangential force in rolling ball spectrometer (see Chapter 9)
G	modulus
G	a parameter in extruder equations (see Chapter 6)
G_0	modulus at zero time
$G(t)$	modulus as a function of time t
G_∞	modulus at infinity time
G'	storage modulus $\Big\}$
G''	loss modulus $\quad$ in dynamic equations
G^*	complex modulus
H	distance between parallel planes
H	logarithmic distribution function of relaxation times
I	current
I	moment of inertia
I_c	capacitative current
I_R	resistive current
J	compliance
J'	storage compliance $\Big\}$
J''	loss compliance $\quad$ in dynamic equations
J^*	complex compliance
J_0	compliance at zero time
$J(t)$	compliance as a function of time t
J_∞	compliance at infinity time
K	a constant
K	specific conductance in ohms^{-1} cm^{-2}

L	length
L_A} L_B}	length of flow unit (C atoms) in polymers A and B respectively
$L(\ln \tau_J)$	logarithmic distribution function of retardation times
M	molecular weight
M	coefficient of thixotropic breakdown with rate of shear
N	Avogadro's number
N	relative velocity of a screw with respect to the barrel (r.p.m.)
P	pressure
ΔP	pressure difference
Q	output (*volume* rate of flow)
Q_m	output (*mass* rate of flow)
R	radius of circular channel
R	gas constant
R	resistance of dielectric in ohms
R	elongation ratio
T	temperature
T_W	temperature at the wall of a channel
T_0	temperature at the centre of a channel
T_0	absolute zero temperature
T_r	reference temperature
V	boundary velocity (sometimes with suffix de- noting axial direction)
V	volume
W	load on ball in rolling ball spectrometer
W	work done
X	a locus
Y	a locus
Y} Y_1} Y_2}	dimensionless quantities in extruder equations (see Chapter 6)
Z	haul-off forces beyond the die exit in extrusion

---

a	a constant
b	a constant
c	a constant
d	complete differential sign
e	base of natural logarithm
$f(\ldots)$	function of ($\ldots$)
$f(\tau_J)$	distribution function of retardation times
$g(\tau_R)$	distribution function of relaxation times

g	geometrical constant (see Chapter 9)
g	gravitational constant
h	height, positional parameter of a point (or plane) between (or parallel to) two parallel planes
$\mathbf{i, j, k}$	unit vectors
k	a constant
k	thermal conductivity
l	length
m	mass
m	reciprocal of power index n
n	power index
$\bar{n}$	mean power index
n	optical refractive index
$\mathbf{q}$	conductive heat flux vector
r	radial distance
r_i	inner radius
r_0	outer radius
s	a function of the power index n (see Chapter 6)
t	time
t_0	zero time
t_∞	infinity time
u_y	dimensionless quantity in extruder equations (see Chapter 6)
v	velocity
v_x, v_y, v_z, etc.	velocity in the direction of a reference axis
v_0	velocity at centre of circular channel
v_f	free volume
v_r	occupied volume
v_l	molar volume in the liquid state
v_s	molar volume in the solid state
$\bar{v}$	mean velocity
$\left.\begin{array}{l} v_x{}^* \\ v_z{}^* \end{array}\right\}$	dimensionless quantities in extruder equations (see Chapter 6)
$\mathbf{v}$	velocity vector
w	slit width
w_y	dimensionless quantity in extruder equations (see Chapter 6)
$\left.\begin{array}{l} x \\ y \\ z \end{array}\right\}$	Cartesian co-ordinate axes

———

α_e	electronic polarisability
α_μ	orientational polarisability
γ	deformation
$\dot{\gamma}$	deformation rate, shear rate

γ_N	Newtonian shear rate
$\ddot{\gamma}$	shear acceleration
$\dot{\gamma}^0$	standard reference shear rate
∂	partial differential sign
δ	loss phase angle
ε'	permittivity (dielectric constant)
ε''	dielectric loss
ε_s	dielectric constant at zero frequency (static)
ε_∞	dielectric constant at infinity frequency
ε_c'	value of ε' when tan δ is at a maximum
ε_c''	value of ε'' when tan δ is at a maximum
η	viscosity
η_N	Newtonian viscosity ⎫ in poise
η^0	standard reference viscosity ⎭
θ	an angle
θ	reference axis in cylindrical and spherical coordinates
ϕ	reference axis in cylindrical and spherical coordinates
μ	permanent dipole moment
ν	kinematic viscosity in Stokes
π	3·14159 · · ·
ρ	density
σ	stress (solid state)
σ_0	stress at zero time
$\sigma(t)$	stress as a function of time t
σ_∞	stress as infinity time
τ	shear stress (liquid state)
τ_y	yield stress
$\boldsymbol{\tau}$	the stress tensor
τ	(with two subscripts denoting axial directions) component of the stress tensor
τ^0	standard reference shear stress
$\boldsymbol{\tau}_a$	anisotropic stress tensor
$\boldsymbol{\tau}_i$	isotropic stress tensor
$(\tau_i)_x$ ⎫ $(\tau_i)_y$ ⎬ $(\tau_i)_z$ ⎭	components of the isotropic stress tensor
τ_J	retardation time
τ_R	relaxation time
ω	angular frequency in radians
ω	resonance frequency

Γ	asymptotic shear rate (see Chapter 1)
Δ	rate-of-strain tensor

Δ	(with two subscripts denoting axial directions) components of the rate-of-strain tensor
∇	'del', 'nabla function', an operator
Λ	logarithmic decrement of damped vibration
Σ	summation sign

Note. Any symbols not listed here are used transiently and are either obvious or defined in the text where appropriate. This applies especially to the Appendix, where the following additional *ad hoc* symbols are used:

$$a, b, A, B, E, F, p, q, k, U, V, W, \Gamma$$

Introduction

RHEOLOGY has been defined as 'a branch of physics which concerns itself with the mechanism of deformable bodies'. Deformation is a phenomenon which is of necessity associated with volume elements. The pioneers of modern rheology concerned themselves principally with the *bulk mechanical* manifestations of the deformational effects of stresses applied to bodies in the liquid state. However, it has become increasingly apparent that rheology cannot be encompassed within such arbitrary confines for the following reasons:

1. The definition of the 'liquid state' is rather arbitrary. A liquid, ideally, is a body which deforms irreversibly as a result of flow. But we know that bodies such as metals—which are indisputably solids—do flow and so deform irreversibly if a force of sufficient magnitude is allowed to act upon them for a sufficient length of time. Flow in solids is known as 'creep'. Structural engineers are well aware of the problems which creep presents. They attempt to cope by designing their structures in such a way as to limit the freedom of relative displacement of the material constituents. They endeavour to 'rigidify' a structure by locking the volume elements so that the design load is insufficient to cause a significant change in the existing spatial arrangement.

The fact that under certain conditions such a change *is* possible implies that at ordinary temperatures there still exists an irreversible flow potential even in such materials as metals. Since the energy required to produce such flow is irrecoverable it must be dissipated as heat. It is true, of course, that an overwhelming proportion of the energy used to deform a piece of metal is instantaneously recovered upon removal of the stress, but it is equally true that some small (and sometimes significant) amount of energy is lost in creep or irreversible flow. The actual percentage of this energy loss depends upon the ease with which the constituent parts of the stressed body can be made to alter their positions in space relative to one another, to slip over one another, to flow.

The resistance of a material to irreversible positional change of its constituent volume elements and the concomitant conversion of mechanical energy into heat is denoted by a parameter known as the 'viscosity' to which we shall assign the correct dimensions in due course. Now, the term viscosity immediately conjures up the concept of a liquid rather than a solid. In order that metals could continue to be regarded as solids, the viscosity of solids is often referred to as the 'internal' or 'frictional' viscosity, as if it were a special kind of viscosity. This devious expertise is both misleading and unnecessary. There is no need for a *qualitative* dividing line between solids and liquids, although the transition from one to the other usually involves a *quantitative* viscosity change which may cover many decades within possibly very close limits of changes in environmental conditions.

The same reasoning can be applied to the gas/liquid transition, since a gas is merely a fluid of particularly low viscosity when compared with a liquid. It is therefore logical to regard the gaseous, liquid and solid states as special aspects of one generalised continuous and universal *fluid state*, with the primary transitions occurring fairly sharply in some materials and less sharply in others. The unifying principle of the universal fluid state lies precisely in the fact that the principal attribute of the liquid state, namely viscosity, exists (obviously) in the gaseous state and (less obviously, but nevertheless demonstrably) in the solid state. This is readily expressed in figures by quoting the order of magnitude of the viscosity of typical materials, viz:

	Viscosity (poise)
Gases	10^{-4}
Liquid air	10^{-3}
Water	10^{-2}
Oils, paints, printing inks	$10^{0}-10^{3}$
Greases, fats	$10^{1}-10^{6}$
Resins	$10^{3}-10^{9}$
Pitches, asphalts, plastics	$10^{6}-10^{12}$
Metals	$\gg 10^{12}$
Ceramics, stone	$\ggg 10^{12}$

2. By inverting the argument, it is also easily seen that materials which are indisputably liquids do not necessarily dissipate *all* the deformational energy. Some of the energy is recoverable and since this is so the liquid has some of the principal attribute of the solid state, namely *elasticity*. Of course, in a typical liquid the magnitude of the viscous response mechanism to an applied stress may be overwhelmingly greater than any manifestation of reversible (recoverable, elastic) deformation.

This can be demonstrated when a high speed ciné film of the impact of a drop of water on a glass plate is examined frame by frame. It is then seen that the drop actually bounces like a ball and returns stored energy (*a*) by rebounding, and (*b*) by recovering its spherical shape after impact instead of maintaining the squat deformed shape which the contact with the glass surface has momentarily imparted to it. We are not concerned at this stage with the nature of the internal or inter-facial forces which manifest themselves by forcing an elastic response. Suffice it to say that liquids are not entirely viscous but possess some of the main attribute of solid state behaviour, in the same way as solids are not entirely elastic but possess *some* of the main attribute of the liquid state.

The best way to describe real materials is to regard them as *viscoelastic*. Some people use two terms (viscoelastic and elasto-viscous) in order to emphasise the predominance of the viscous or elastic response respectively, but the writer feels that this distinction is somewhat pedantic.

3. Rheology cannot allow itself to be restricted only to bulk mechanical deformations of the various aspects of the fluid state. Since volume elements are involved in *all* deforma-tions, whatever the force field, there is no reason why the mathematical equations developed to describe mechanical deformations should not basically apply to deformations induced by an electrical, magnetic or any other force field. Volume elements will be different in each case, to be sure; in a plastic under tension or in melt flow, viscosity manifests itself by the spatial rearrangement of polymer chain segments which constitute quite large volume elements. An electric field acting on polar plastics will cause energy dissipation as frictional heat when the dipoles do work against their environ-ment in their endeavour to conform to the polarity of the applied field. The volume element is smaller, but the same argument applies.

Again, excitation by light rays (visible or otherwise) will cause certain deformations of which some will be irreversible. In the infra-red the volume elements involved are specific atomic bond conformations and the energy dissipation is implicit in the frictional heat generated when these bonds are partially constrained in their vibrational and rotational evolutions by their environment. The volume elements are perfectly real, although they require a highly specific mode of excitation before they can manifest themselves.

Similar arguments can be applied to nuclear magnetic and electron-spin resonance where the volume elements are subatomic but no less real for that. We cannot therefore content ourselves with a statement that rheology is 'a branch of physics'. It is far more than that. It is the key to the under-standing of the behaviour of any substance which is subjected to any kind of force field.

This casts the net rather widely. Indeed, far from being a branch of any science, rheology emerges as *the* central science with such branches as chemistry, physics, engineering, biology, etc. In their innate generosity, rheologists are quite content to lease part of their territory to the devotees of these marginal disciplines: we do not concern ourselves overmuch with turbulent flow except when it suits us (as in polymer processing or lubrication problems). We cheerfully leave it to the aerodynamicists to sort out the flow lines of air around flying machines—but we keep a watchful eye on gaseous disperse systems such as fluidised beds. We are not primarily concerned with establishing the precise nature of either the forces which cause resistance to deformation or the volume elements on which the applied force field acts and leave it to the physicist and physical chemist to find out. But while we may not be involved in establishing some of the facts we do wish to know these facts and are duly grateful to those who work in auxiliary disciplines in order to provide us with the necessary data on which to buttress the crowning glory of an all-embracing rheological superstructure.

Having discoursed at some length on rheology in general we shall now have to turn to plastics rheology in particular.

Rheologists work either on a purely mathematical basis without bothering overmuch about specific materials, or they study a limited range of materials on a phenomenological basis, designing experiments which they hope may subsequently be capable of yielding at least empirical (but preferably theoretically justifiable) generalisations.

It is often said that the mathematical and experimental rheologists are two races apart—and there is more than a grain of truth in this, however regrettable this may be. The mathematics involved are often difficult and there is therefore a lack of communication in evidence. This state of affairs will doubtless mend itself in time, as integrated rheologists begin to roll off the production lines of higher educational technological establishments. Indeed, in the case of plastics rheology there already exists a fair amount of integration due to the work of a large number of eminent scientists over the last thirty years, and if this book succeeds in making a modest contribution to a more conscious and purposeful integration it will have fulfilled its main purpose.

What has made plastics rheology a particularly fascinating branch of general rheology is the fact that plastics exhibit a spectrum of deformational responses to stresses which concerns itself in particular with the border region between the 'solid' and 'liquid' states and uses this platform as a springboard for the further penetration of the more typical regions of these two states. In addition, the industry has discovered the fundamental importance of studying plastics by methods which had to be specially developed, since the methods which were acceptable for traditional materials proved to be totally inadequate if a full understanding of the formulation, tailoring, compounding, processing, design and functional performances of plastics is to be gained. What is more, this realisation has caused the plastics chemist, plastics physicist, plastics technologist and plastics engineer to become rheology-conscious.

There are few fields in science in which the specialist can be persuaded to leave his ivory tower even momentarily. But the field of plastics is undoubtedly one of them and it is due to this readiness to survey the wider field and his specific role within it that the plastics scientist has been able to make a more effective contribution to the development of his industry than would otherwise have been possible.

In addition, there has been a most welcome realisation on the part of the theoreticians that here exists a field in which their mathematics can be applied and put to the test. This has led to important advances especially in the extrusion and calendering processes and here Pearson's work in the United Kingdom and McKelvey's, Carley's and Merz's work in the U.S.A. deserve special mention. Vincent, Turner, Atkinson and their colleagues have succeeded in drawing attention to the ultimate importance of service performance realities and they have established criteria for the correct assessment of the strength of plastics in the solid state.

Rheology has made important contributions to advances in food technology, medicine, paint and printing-ink technology, building and structural engineering, adhesives, cosmetics, oilwell drilling operations, and elsewhere. But it cannot be denied that it is due to its tremendous scope in plastics that rheology in general has now reached a status of maturity leading to its recognition as the major scientific discipline of tomorrow. If rheology were to nail a motto to its mast it could do worse than choose the wise words of old Thales of Miletus: 'Everything flows'.

1 Flow in the liquid state

Types of deformation (general). Viscous deformation and its definition. Derivation of the Hagen-Poiseuille Equation as a special case of a general power law. The Rabinowitsch Correction. Non-Newtonian materials. Bingham body, pseudoplasticity, dilatancy, first and second Newtonian regions. Molecular interpretation of changes in flow regime. Time dependency—thixotropy and rheopexy, discussion of experimentally obtained flow curves and artefacts (cavitation, slippage). The generalised flow curve. Elasticity (normal forces) in liquids and the molecular basis of this phenomenon. The theoretical basis of the power law.

WE have established that deformation is of two kinds:

1. Spontaneously reversible (elasticity).
2. Irreversible (flow).

Concerning ourselves first with flow it is important to note (1) that energy is continuously required to sustain flow and (2) that this work is not mechanically recoverable but is dissipated as heat. The resistance of a fluid to flow is its viscosity or its 'internal friction'.

An important type of flow deformation is *shear*. Simple shear between parallel plates can be considered as a process in which infinitely thin parallel planes slide over each other 'like a pack of cards'. Simple shear between parallel plates is a special case of laminar deformation in which the laminae are flat planes, but laminar flow can also be found in other geometries. Of those other geometries the cylindrical is of the greatest importance, as shown, for example, in Fig. 1.1.

Rotational shear is especially important in cup/bob type and twisting shear in cone/plate type rotational viscometers, whilst telescopic shear is especially important in capillary and extrusion rheometers.

Simple shear between parallel plates and twisting shear are commonly used in determining the shear modulus of solids.

In rotational shear, deformation is the process in which the relative change in position of volume elements occurs without changing the overall volume or shape. The deformation which results in a change of shape is called *distortion*. Distortion may be recoverable (as in a spring) or irreversible (as in a lump of putty).

Elastic deformation at constant environmental conditions is a function of stress. Viscous deformation is likewise a function of stress, but since it manifests itself by flow, it is necessary to maintain that flow in order to maintain the stress. This can only be achieved by continuous agitation of the liquid. The stress in a liquid is therefore, in turn, a function of the *rate*

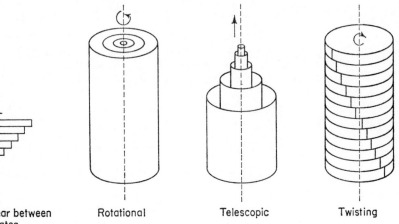

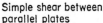

| Simple shear between parallel plates | Rotational | Telescopic | Twisting |

Fig. 1.1

of agitation, the *rate* of deformation, the *shear rate*. This leads to an important realisation: Deformation in the solid state, expressed by the ratio of elongation (in a tensile experiment) and the original specimen length is a *dimensionless quantity*, $\gamma = \mathrm{d}l/l$, whilst the shear rate in a liquid is a *time derivative*, $(\mathrm{d}l/l)/\mathrm{d}t$. It will be noted that this expression can also be written $(\mathrm{d}l/\mathrm{d}t)/l$, that is to say, velocity per unit length, where $\mathrm{d}l$ and l have, of course, the same units, commonly denoted by the symbol $\dot{\gamma}$. The viscous deformational analogue (shear rate) which applies to liquids and corresponds to elastic deformation (elongation ratio) in the solid therefore differs from the latter dimensionally. Elastic deformation is expressed in terms of a dimensionless ratio; viscous deformation (flow) is expressed in terms of the *shear rate* the unit of which is reciprocal time.

The differential equation which describes viscous deformation will, then, involve the *rate* of deformation, the shear rate. If the deformation is not *wholly* viscous, as is invariably the case in real bodies, then the differential equation which describes the overall deformation requires an additional term which involves elastic strain. Even then the differential equation is not completely rigorous, since it still lacks an inertial term involving acceleration. The shear rate observed in viscometry is that which operates at the wall of the instrument. This shear rate is not typical of the total volume of fluid being sheared and its representative value depends on the boundary conditions. A better representation of the shear rate which may be assigned to the volume under shear would be an average value. To obtain this, one requires a knowledge of the distribution of shear rates throughout the volume.

To produce flow, a stress must be applied. The stress is defined as a force per unit area. Since a liquid will tend to respond to an applied stress by escaping tangentially because of the relatively large influence of the boundary forces, the stress will be shear rather than tensile or compressive. The force necessary to produce flow in a liquid must be of such a magnitude as to overcome the forces of attraction between molecules, so that the molecules are displaced relative to one another. The stronger the inter-molecular forces, the smaller is the amount of flow for a given applied force and the greater is the viscosity of the liquid.

The basis of viscosity measurement was laid by Newton who postulated that the rate of flow is proportional to the applied stress, the proportionality constant being the coefficient of viscosity η:

$$\eta\dot{\gamma} = \tau$$

where

$$\dot{\gamma} = \text{rate of shear}$$
$$\tau = \text{shear stress}$$

This relationship is formally identical (except for the time dependence of the rate of shear) with any generalised stress-strain relationship and can therefore be regarded as a special type of modulus applicable to liquids.

The Newtonian model consists of a stationary plane and a moving plane parallel to the stationary plane and separated by a distance r, with the liquid occupying the volume between them and moving in laminar fashion in the direction of the moving plane, but with a velocity which is a linear function of the distance from the moving plane. The shear stress applied to the moving plane has the dimensions of force per unit area (dynes cm^{-2}) and the rate of shear at any point between the planes is given by $\dot{\gamma} = dv/dr$, where the velocity v has the units of cm sec^{-1} and r is in cm. The units of $\dot{\gamma}$ are therefore sec^{-1} ('reciprocal seconds') and the viscosity η is defined as the tangential shearing force per unit area which will produce a unit velocity gradient: $\tau = \eta\,(dv/dr)$, where the shear stress τ is the tangential shearing force per unit area in dynes cm^{-2}. The viscosity may therefore be determined either by measuring

3

the rate of shear caused by a known shear stress, or by measuring the shear stress required to produce a known rate of shear.

In practice it is frequently required to express the output of liquid flowing through a channel in terms of the applied pressure, rather than shear stress in terms of shear rate or vice versa. If it is assumed that the Newtonian relationship $\tau = \eta\dot{\gamma}$ holds, then a simple output/pressure drop equation can be derived. This equation is known as the Hagen-Poiseuille equation.

However, the assumption of direct proportionality between τ and $\dot{\gamma}$ is correct only at low (and in the case of polymer melts at vanishingly low) shear rates. Under processing conditions a 'power law' takes the place of the direct Newtonian relationship: $\tau = \eta\dot{\gamma}^n$, where n is the 'power index'. It is evident that for the special case of $n = 1$ the special Newtonian case is recovered from the more general power law. In the following the output/pressure drop equation for a power-law liquid flowing through a circular channel is derived and by setting $n = 1$ in that equation the Hagen-Poiseuille equation is obtained.

If the shear rate at the wall of the channel of a power-law liquid at vanishingly low shear stresses (when $n = 1$ and the flow is Newtonian) is known, and if the rate of change of the shear rate at the wall with shear stress $(d\dot{\gamma}/d\tau)$ in the power law region is also known, then it is possible to calculate the shear rate at the wall for any shear stress within the power-law flow regime. The equation expressing the shear rate at the wall in the power-law regime in terms of the two parameters named at the head of this paragraph is known as the 'Rabinowitsch correction'. The derivation of the Hagen-Poiseuille equation and the Rabinowitsch correction is now given.

Derivation of the Hagen-Poiseuille equation and the Rabinowitsch correction

Consider the balance of forces for flow through a capillary:

$$\tau \quad \times \quad 2\pi RL \quad = \quad \Delta P \quad \times \quad \pi R^2$$

shear stress per unit area	area of contact of moving cylinder of liquid along capillary wall	pressure difference at capillary ends	area over which pressure is exerted

Total shear stress   Total work

whence

Equation 1.1
$$\tau = \frac{\Delta PR}{2L}$$

In a power-law liquid

Equation 1.2
$$\tau = \eta\dot{\gamma}^n = \frac{\Delta Pr}{2L} = \eta_N\left(\frac{dv}{dr}\right)^n$$

or

$$\frac{dv}{dr} = \left(\frac{\Delta Pr}{2L\eta_N}\right)^{1/n}$$

which on integration* gives:

$$v(r) = -\frac{n}{n+1}\left(\frac{\Delta P}{2\eta_N L}\right)^{1/n} r^{(n+1)/n} + C$$

The integration constant is evaluated from the boundary condition that $v(r) = 0$ at $R = r$, whence

Equation 1.3
$$v(r) = -\frac{n}{n+1}\left(\frac{\Delta P}{2\eta_N L}\right)^{1/n} R^{(n+1)/n}\left[1 - \left(\frac{r}{R}\right)^{(n+1)/n}\right]$$

The velocity at the centre, v_{max}, is obtained by setting $r = 0$, so that

Equation 1.4
$$v_{max} = -\frac{n}{n+1}\left(\frac{\Delta P}{2\eta_N L}\right)^{1/n} R^{(n+1)/n}$$

or

Equation 1.4
$$\frac{v(r)}{v_{max}} = 1 - \left(\frac{r}{R}\right)^{(n+1)/n}$$

also known as the 'reduced velocity profile'. Note that for a Newtonian, where $n = 1$, the power of r/R becomes 2 and for $r = 0$, i.e. at the centre, $v(r) = v_{max}$ always.

The mean velocity $\bar{v}$ is calculated by summing all the cross-sectional velocities and dividing by the cross-sectional area. The output Q is given by the sum of the velocity in each cross-sectional area element multiplied by the area of each cross-sectional area element.

* Integration between limits does away with the constant and yields the same result:

$$v(r) = \left(\frac{\Delta P}{2L\eta}\right)^{1/n}\int_r^R r^{1/n} = \left(\frac{\Delta P}{2L\eta}\right)^{1/n}\frac{r^{1/n+1}}{\frac{1}{n}+1}\bigg|_r^R = \left(\frac{R^{1/n+1}}{\frac{1}{n}+1} - \frac{r^{1/n+1}}{\frac{1}{n}+1}\right)$$

$$= \left(\frac{\Delta P}{2L\eta}\right)\frac{R^{(n+1)/n} - r^{(n+1)/n}}{\frac{n+1}{n}} = \frac{n}{n+1}\left(\frac{\Delta P}{2L\eta}\right)^{1/n}R^{(n+1)/n}\left[1 - \frac{r^{(n+1)/n}}{R}\right]$$

identical with (1.3).
(The negative sign only indicates the direction of flow and can be neglected *if* the correct sign is given to the pressure drop.)

In the limit this becomes:

$$Q = \int_0^R v(r) \,.\, 2\pi r \,dr$$

where $v(r)$ is given by Equation 1.3. Substituting accordingly and putting

Equation 1.5
$$\left[\left(\frac{\Delta P}{2L\eta_N}\right)^{1/n} \frac{n}{n+1}\right] = a$$

$$Q = 2\pi a \int_0^R \left(R^{(n+1)/n}\, r - r^{(n+1)/n}r\right)\,dr$$

$$= 2\pi a \int_0^R \left(R^{(n+1)/n}\, r - r^{(2n+1)/n}\right)\,dr$$

Integrating,

$$Q = 2\pi a \left(R^{(n+1)/n}\frac{r^2}{2} - \frac{r^{(3n+1)/n}}{\dfrac{3n+1}{n}}\right)\Bigg|_0^R$$

Taking the limits, this becomes:

$$Q = 2\pi a\left(\frac{R^{(3n+1)/n}}{2} - \frac{R^{(3n+1)/n}}{\dfrac{3n+1}{n}}\right) + \pi a\,\frac{\dfrac{3n+1}{n}R^{(3n+1)/n} - 2R^{(3n+1)/n}}{2\cdot\dfrac{3n+1}{n}}$$

Resubstituting for a according to Equation 1.5:

$$Q = \left[\pi \left(\frac{\Delta P}{2\eta_N L}\right)^{1/n} \frac{n}{n+1}\right]\frac{\left(\dfrac{3n+1}{n} - 2\right) R^{(3n+1)/n}}{\dfrac{3n+1}{n}}$$

$$= \left[\quad\right]\frac{\dfrac{n+1}{n}R^{(3n+1)/n}}{\dfrac{3n+1}{n}}$$

Equation 1.6
$$= \pi\left(\frac{\Delta P}{2\eta L}\right)^{1/n}\frac{n}{3n+1}R^{(3n+1)/n}$$

Dividing the output by πR^2, the cross-sectional area, we obtain the mean velocity $\bar{v}$ (by definition):

$$\bar{v} = \frac{Q}{\pi R^2} = \left(\frac{\Delta P}{2\eta_N L}\right)^{1/n}\frac{n}{3n+1}R^{(3n+1)/n-2}$$

Equation 1.7
$$= \left(\frac{\Delta P}{2\eta_N L}\right)^{1/n}\frac{n}{3n+1}R^{(n+1/n)}$$

For a Newtonian ($n = 1$) (1.7) becomes:

$$\bar{v} = \frac{\Delta P}{2\eta_N L} \cdot \frac{1}{4} R^2 = \frac{\Delta P}{8\eta_N L} R^2$$

whereas, according to (1.4),

$$v_{max} = \frac{\Delta P}{4\eta_N L} R^2$$

Hence *in a Newtonian*,

$\bar{v} = \frac{1}{2}v_{max}$, whilst in a general power-law liquid

Equation 1.8
$$\bar{v} = \frac{n + 1}{3n + 1} v_{max}$$

Since for a Newtonian,

$$\bar{v} = \frac{\Delta P}{8\eta_N L} R^2 \left(= \frac{Q}{\pi R^2} \text{ by definition of } \bar{v} \right)$$

$$\frac{4Q}{\pi R^3} = \frac{\Delta P R}{2\eta_N L} \equiv \tau,$$

the shear stress at the wall, where $r = R$.
Rewriting the above equation,

Equation 1.9
$$Q = \frac{\pi \Delta P R^4}{8\eta_N L} \text{ (the Hagen-Poiseuille equation)}$$

Making η_N the subject of this equation,

$$\eta_N = \frac{\pi \Delta P R^4}{8QL} = \frac{\tau}{\dot{\gamma}_N} \text{ (by definition)}$$

$$\therefore \quad \frac{\Delta \pi P R^4}{8QL} = \frac{\Delta P R}{2\eta_N L \dot{\gamma}_N} \text{ (by substitution for } \tau)$$

Equation 1.10
$$\therefore \quad \dot{\gamma}_N = \frac{4Q}{\pi R^3}$$

Substituting for $4Q/\pi R^3$ in the general power-law equation (i.e. writing $\dot{\gamma}_N/4$ for $Q/\pi R^3$), and remembering that

$$Q = \pi R^2 \cdot \bar{v} = \pi R^2 \cdot \frac{n + 1}{3n + 1} \cdot v_{max} =$$

$$\uparrow \qquad\qquad\qquad \uparrow$$
by definition $\qquad\quad$ by 1·8

$$= \pi R^3 \frac{n}{3n + 1} \left(\frac{\Delta P R}{2\eta_N L} \right)^{1/n}, \text{ (using } 1·4)$$

$$\uparrow$$

The bracket term is recognised as the shear rate $\dot{\gamma}$ at the wall,

7

so that
$$Q = \dot{\gamma} \frac{n}{3n + 1} \pi R^3$$

$$\therefore \dot{\gamma} = \frac{3n + 1}{n} \frac{Q}{\pi R^3} = \frac{3n + 1}{n} \frac{\dot{\gamma}_N}{4}$$

Equation 1.11
$$= \frac{\dot{\gamma}_N}{4} \left(3 + \frac{1}{n} \right) = \frac{\dot{\gamma}_N}{4} \left(3 + \frac{d \log \dot{\gamma}}{d \log \tau} \right)$$

by definition of n. Obviously, in a Newtonian where $n = 1$

$$\dot{\gamma} \equiv \dot{\gamma}_N$$

This is the *Rabinowitsch correction* which enables one to calculate the shear rate at the wall of a power-law liquid if the shear rate $\dot{\gamma}_N$ in the Newtonian region and the power index n are known.

The absolute (cgs) unit of viscosity is the poise, so named in honour of Poiseuille who formulated the relationships of shear stress and shear rate in circular channels. Its dimensions from Newton's basic equation are $ml^{-1}t^{-1}$.

Another unit of viscosity is the 'stoke' (v) with dimensions l^2t^{-1}. This refers to the 'kinematic viscosity'.

Dimensional analysis of the poise and the stoke establishes the relationship

$$v = \frac{\eta}{\rho}$$

where ρ, the density in g/cc has the dimensions of ml^{-3}.

Kinematic viscosity is used with viscometers in which the measurement is influenced by the density of the liquid, for instance in gravity operated capillary viscometers.

The reciprocal of viscosity is known as the fluidity (or the ease of flow) and is expressed in rhes. Fluidity bears the same relationship to the viscosity of liquids as compliance bears to the modulus in solids.

Whilst viscosity determines the degree of flow, observation has shown that viscosity is not necessarily constant even at constant temperature. This is contrary to Newton's postulates. Whilst Newtonian behaviour is common in many simple liquids under ordinary experimental conditions, it is restricted to certain regions of shear stress in more complex but nevertheless commonly occurring systems, especially when the viscosity is high. Amongst systems which show marked deviation from the Newtonian shear stress/shear rate relationship are the following:

1. Systems in which a disperse phase constitutes a substantial volume fraction, such as paints, starch/clay paper coating mixes, toothpaste, mustard, custard, blood, oilwell drilling muds, molten chocolate, printing inks, fluidised beds, adhesive pastes, fibre slurries, dough, polymer emulsions, plastisols, etc.

2. Systems in which the unit of flow can attain an orientated condition from a random ground state under the influence of an applied shear field and in which—as distinct from (1)—no macroscopic inhomogeneity is apparent, although inhomogeneity on another level may well be present. Such systems include high viscosity oils, polymer solutions and polymer melts.

Furthermore, deviation from Newtonian behaviour may not only be shown in the shear rate dependency of viscosity, but also in changes of viscosity at constant shear rate and temperature as a function of time. It is well known that certain systems on continued isothermal shearing at constant shear rate show a progressive reduction in viscosity to some limiting value. This is due to a reversible breakdown of such 'structural' features as may be present in the random ground state, coupled with the fact that the process of rebuilding these 'structural' features has a lower rate constant than the process of breakdown. When the rate constants of the two processes differ sufficiently so as to become experimentally observable we say that the liquid is 'time dependent'. It is clear that time dependence cannot exist under a Newtonian regime where there is no 'structure' present which could be broken down. When time dependence is observably superimposed on the shear stress/shear rate relationship we must then state the previous shear history of the specimen in order to make viscosity figures meaningful. Observable time dependence is not common in polymer melts and the processing of polymers which, of course, involves a transient melt stage is not generally complicated by time dependence.

It should be clear from what has been said that no *one* instrument can give a comprehensive picture of the flow behaviour of a polymer material, since a wide spectrum of shear rates needs to be investigated over wide temperature ranges and possibly over varying periods of time, and this may involve viscosity changes of several decades.

For a comprehensive survey of commercially available instruments the reader is referred to Van Wazer *et al.*,* where the construction and operation of instruments is described in detail.

---

* Van Wazer, Lyons, Kim, and Colwell, *Viscosity and Flow Measurements*, Interscience (1963).

Non-Newtonian liquids may be classified as follows:

1. Liquids in which the shear stress depends on the shear rate alone.

2. Liquids in which the relationship between shear rate and shear stress depends on the time of shear, including that part of its shear history which precedes the experimental investigation.

3. Liquids in which the instantaneous elastic component of shear deformation becomes significantly large and can no longer be ignored, even as a first approximation. (Many workers nowadays incline to the view that most liquids fall into that group and so far as polymer melts are concerned, partial recovery or stored elasticity is very much in evidence in the extrusion and calendering processes.)

1. Liquids in which the shear stress depends on shear rate alone

Assuming these to be entirely time-dependent and essentially devoid of elasticity under the conditions of test, these show the following possible deviations from Newtonian behaviour:

(a) No deformation occurs until a certain threshold shear stress is applied, whereupon the two parameters are in linear relationship, that is to say, the shear stress τ becomes a linear function of shear rate $\dot{\gamma}$. The characteristics of the function are the slope (viscosity) and the shear stress intercept ('yield value') τ_y. If the rheological equation for a Newtonian is $\tau = \eta\dot{\gamma}$, then the rheological equation for this type of material, known as a 'Bingham body' is:

$$\tau - \tau_y = \eta\dot{\gamma}$$

Since we cannot have either negative shear rate or negative viscosity, this equation only makes sense for values of $\tau \geqslant \tau_y$. Only when the applied stress is greater than the yield stress does viscous flow and shear deformation become manifest.

Substances which exhibit this type of behaviour include oil-well drilling muds, sewage sludge, toothpaste, greases and fats, but since it excludes plastics we need not concern ourselves with it any further.

(b) Shear stress is not a linear but an exponential function of shear rate. This represents the common power law behaviour of polymer melts, the rheological equation of which is

$$\tau = \eta\dot{\gamma}^n$$

Here we can recognise three different situations:

(i) Clearly, if $n = 1$, then the equation reduces to the Newtonian case.

(ii) If $n > 1$, then, for any increase in shear rate the shear stress increases at a *greater* rate. The curve which was linear for $n = 1$ now becomes *convex* with respect to the shear stress axis. Such a material is known as 'dilatant'.

A dilatant material is so called because of the observation that the $\tau/\dot{\gamma}$ relationship which exists in this instance may be accompanied by an expansion in volume. This state of affairs is found in disperse systems when the disperse phase is very

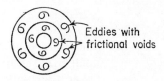

Eddies with
frictional voids

Low shear rate,
Newtonian regime,
perfect laminae

High shear rate,
Dilatant regime,
interlaminar eddies formed
radius of volume element increased

Fig. 1.2

crowded, so that the corresponding scarcity of continuous phase makes it difficult for the disperse particles to slide over one another. The lack of adequate lubrication by the continuous phase due to its quantitative insufficiency will produce frictional voids and set up 'structures' amongst the particles under constraint. These 'structures' become increasingly resistant to deformation as the applied shear rate is increased: the viscosity increases with increasing shear rate. This behaviour is seen in oil paints which have settled into a bottom cake of pigment. When, after decanting the supernatant continuous phase, an attempt is made to stir the sediment it becomes obvious that 'the more one tries to stir it, the more it does not flow.'

Another explanation for the appearance of dilatancy which the writer suggests and which is especially attractive for polymeric as well as disperse systems is as follows:

During Newtonian flow the shear laminae do not interfere with one another to any observable degree, but as the shear rate is increased neighbouring laminae cause eddies to form. This constitutes 'structural' build-up and must therefore result in a change from a Newtonian to a dilatant flow regime under which the resistance to flow (i.e. the viscosity) increases, as shown in Fig. 1.2.

It is easily seen that at some critical high shear rate the 'balling-up' process will progress to such an extent and the structural build-up will become so great that the flow literally grinds to a halt, the viscosity will tend to infinity and the liquid will assume the attributes of an elastic solid. In polymer melts the chains, originally—that is to say at low shear rates—aligned along the shear lines, would 'ball-up'. This phenomenon is, of course, reversible and on resting (or on reducing the shear rate to a suitably low level) the entanglements and eddies will return to the ground state, whereupon the material will re-assume its typical liquid appearance.

Recent work of van der Vegt and Smit [1] has provided powerful support for this view. They have found extreme dilatancy in polymer melts above certain shear rates and they proved by X-ray crystallography of the quenched melt that the melt had attained a high degree of crystallinity. That the 'balled up' volume elements previously described should in fact turn out to be crystallites created by shear-induced orientation in the melt, just as rapidly strained rubbers may do in the 'solid' state, does not invalidate the conception in the slightest—in fact it makes it more precise. Together with Galt and Maxwell's [2. v. 150] work it provides evidence for the existence of dilatancy in many materials hitherto regarded as being characterised by a different flow regime and so proves the correctness of the concept of a generalised flow curve which is developed later in this chapter (pp. 22–25).

(iii) If $n < 1$, then the reverse situation exists: The $\tau/\dot{\gamma}$ curve is now *concave* with respect to the shear stress axis. This may be explained by considering the shear stress as a force which progressively breaks down the random ground state 'structure' to an extent determined by the shear rate. If that breakdown process requires a minimum shear rate before it becomes observable, then the material will flow in Newtonian fashion below that minimum shear rate. The change of flow regime from the low-shear-rate Newtonian region at that critical point which marks the onset of deviation from the Newtonian behaviour is known as the lower threshold of 'pseudoplastic' behaviour. At that point the power index n becomes less than unity and the material responds according to Le Chatelier's principle which states that 'when a constraint is exercised upon a system, the system will adjust itself so as to conform to that constraint.' The mode of adjustment in a pseudoplastic liquid, clearly consists of alignment along shear planes in such a way as to reduce frictional resistance. This involves the progressive breakdown of the random ground state 'structure' and build-up of a shear orientated 'structure' and therefore easier flow, that is to say, a reduction in viscosity. In polymer melts the initial Newtonian region is evident only at very low shear rates indeed and may be easily missed when taking flow measurements, but it is obvious that the structural breakdown process cannot occur until some critical shear rate threshold—however low this may be—has been exceeded.

Materials in which pseudoplasticity is readily observed include paints, printing inks and polymer melts.

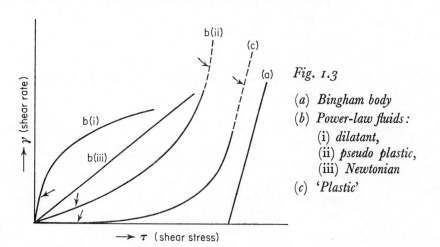

Fig. 1.3

(a) *Bingham body*
(b) *Power-law fluids:*
   (i) *dilatant,*
   (ii) *pseudo plastic,*
   (iii) *Newtonian*
(c) *'Plastic'*

The question now poses itself as to what may be expected to happen if the shear rate has been increased to such an extent that all residual random ground state 'structure' has disappeared and the flow units are all aligned in laminar fashion along the shear planes. Clearly, any further increase in shear rate cannot shear out any more 'structure', because all 'structure' that can be sheared out *has* been sheared out. The viscosity must therefore remain constant from there on—in other words, we are entering a second region of Newtonian flow regime. In practice this is not easily demonstrated with polymer melts, because the high shear rates required, coupled with the (still) very high viscosity of polymer melts under those conditions, result in laminar flow being superseded by turbulence or other gross flow irregularites. But the existence of a second Newtonian region has long been recognised in polymer solutions and in disperse systems. We shall return to this in due course.

The $\dot{\gamma}/\tau$ relationship is shown in Fig. 1.3 above. Curve (a) applies to a Bingham body, curve b (i) to a dilatant power law liquid, curve b (ii) to a pseudoplastic power law liquid and curve b (iii) to a power law liquid in which the power index n is exactly unity and which therefore behaves in a 'Newtonian' fashion.

There also exists a hybrid condition which is found in some liquids but which has not been demonstrated with certainty to exist in polymer melts—curve (c), in which a pseudoplastic curve has been shifted along the shear stress axis from the origin and which therefore bears the same relationship to the pseudoplastic as the Bingham body bears to the Newtonian. This behaviour is called 'plastic'. An initial Newtonian portion in a 'plastic' curve has never been observed with certainty and it is obvious that the initial Newtonian portion is in fact represented by the abscissa itself. This means that up to a point the liquid has 'infinity viscosity', a statement which, despite its apparent paradoxical nature is nevertheless entirely logical.

Note the linearity of the arrowed parts of the curves which refer to the low shear rate and (dotted) high shear rate Newtonian regime regions.

2. Liquids in which the $\dot{\gamma}/\tau$ relationship depends on the time of shearing

Strictly speaking, all liquids which exhibit build-up or breakdown of 'structure' with increasing shear rate are time dependent in the sense that the process of structural change occurs over a finite time interval.

Let us apply a uniformly increasing shear rate (that is to say, a constant shear acceleration*) to a pseudoplastic up to some arbitrary top shear rate beyond the initial Newtonian region. Immediately on reaching the top shear rate let us then apply an equally uniformly decreasing shear rate (shear deceleration), returning to the rest position over the same time interval.

It does not follow of necessity that the rate of structural build-up during shear deceleration must equal that of structural breakdown during shear acceleration. If the structural changes are both very rapid, then the upcurve and downcurve will coincide and will not be experimentally separable. If, however, the rebuilding process is very much slower than both the breakdown process and the timescale of the experiment, then one would expect the two curves to be non-coincident. The resulting hysteresis loop will be characteristic for the material at any stated shear acceleration and top shear rate. This characteristic behaviour is known as *thixotropy* and is commonly found in disperse systems (but not commonly in polymer melts). In many cases the rebuilding process is extremely slow and may require minutes, hours, days or even weeks at rest before it is completed.

* Shear acceleration is the second time derivative of deformation. It is therefore symbolised by $\ddot{\gamma}$.

In cases of extremely slow rebuilding the sheared material will give not only a typically linear Newtonian downcurve, Fig. 1.4 *d* (i), but will furthermore follow that same straight line if the material is repeatedly put through the same cycle without allowing it to rest and regain the random ground state 'structure', Fig. 1.4 *d* (ii). Such a material may therefore be mistaken for a Newtonian liquid. Most common thixotropic materials, notably oil paints, recover *part* of their structure on the downward lap of the hysteresis loop and this manifests itself in a curvature towards the shear stress axis. This is most clearly seen when the curve passes through the low shear rate region which is arrowed in Fig 1.4 (*c*).

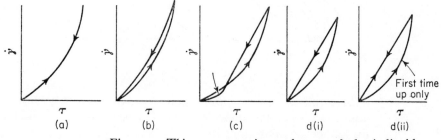

Fig. 1.4. Thixotropy superimposed on pseudoplastic liquids.

(*a*) *No macroscopically observable thixotropy—build-up after break-down virtually instantaneous on reduction of shear rate.*
(*b*) *Some thixotropy apparent—build-up after breakdown fast, but at an experimentally observable rate.*
(*c*) *More thixotropy than in (b), but build-up observable in the arrowed lower shear rate region.*
(*d*) (i) *Highly thixotropic. The material must be rested before the upcurve can be exactly reproduced.*
(*d*) (ii) *Highly thixotropic. Build-up is so slow that an early repetition of the up-and-down cycle does not permit sufficient resting time for the reattainment of the original random ground state 'structure'. Consequently, the downcurve is reproduced on both up and down laps of subsequent cycles.*

Hysteresis loops can also appear when a *dilatant* liquid loses the 'structure' which has been built up during the upcurve. The rate at which 'structure' is lost during the downcurve depends on the shear deceleration during the upcurve. If the upcurve was obtained at a high shear acceleration only part of the potentially available 'structure' is realised. The hysteresis loop appears because, at any given shear rate, there is more 'structure' present on the downcurve than on the upcurve, although (on an absolute scale), 'structure' may be decreasing progressively on the downcurve itself (Fig. 1.5, (b), (c), (d)). In some cases, however, the build-up process may actually continue for part of the downcurve (e, f) and go through a maximum, whereupon fast breakdown occurs; but in extreme cases the build-up will continue all the way down to almost zero shear rate and only after complete rest for long periods of time will the shear stress gradually decay (g).

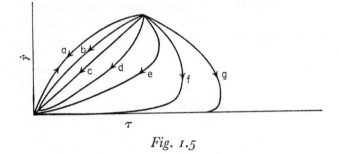

Fig. 1.5

Materials of type (e) and (f) have been observed with certain aqueous starch/clay mixes of a kind which are utterly useless for paper coating purposes. The shapes of the curves are entirely determined by the coefficients of the different rate processes. If the upcurve is made so slowly (i.e. if the shear acceleration is so low) that the full 'structure' building potential is realised at any given shear rate, then it is of course impossible for more structure to be built up on the downcurve and the curves in Fig. 1.3 above would be as shown in Fig. 1.4:

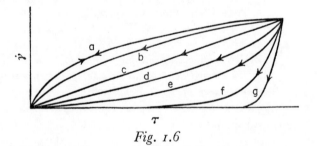

Fig. 1.6

The behaviour which brings about a hysteresis loop in *dilatant* materials is referred to as 'rheopexy'.

It will be noted that the thixotropic and rheopectic hysteresis loops run anti-clockwise and clockwise respectively (shear stress plotted as abscissa).

The rate at which 'structural' breakdown (pseudoplastics) or build-up (dilatants) occurs will determine the extent to which a material will respond to changes in shear acceleration. The slower the return build-up or return breakdown the more pronounced will be the observed degree of thixotropy or rheopexy (respectively). *In addition*, the *ratio* of the two rate constants for build-up and breakdown is also a function of shear acceleration.

Green and Weltmann attempted to separate the two for thixotropic materials and defined coefficients of thixotropic breakdown:

(*a*) with rate of shear at constant shear acceleration,

(*b*) with time at constant rate of shear,

but with the concept of shear acceleration borne in mind this distinction would not seem to be fundamental, although it does have practical uses in characterising materials.

Certainly, a thixotropic liquid possesses a structure which breaks down until a steady state is reached at which build-up and breakdown occur at equal rates. This is shown when the liquid is taken from zero shear rate to some arbitrary top shear rate as quickly as possible and maintained at that shear rate. The shear stress (at a high level when top shear rate is reached) will decay monotonically to a constant value. Viscosity decreases in proportion to shear stress. The viscosity can therefore have any value between that first observed (cotangent at point A) and that eventually attained (cotangent at point B) as shown in Fig. 1.7, where the same liquid was brought to an arbitrary top shear rate at shear accelerations which increase from left to right.

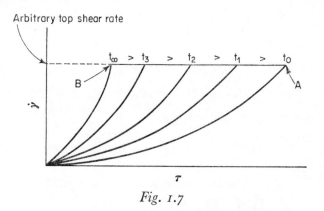

Fig. 1.7

The downcurve will commence at any point between A and B, depending only on how long the liquid was left to shear at top shear rate. Once point B has been reached, however, no further viscosity decrease is possible except on increasing the shear rate.

Curves $\dot{\gamma}_1$, $\dot{\gamma}_2$ and $\dot{\gamma}_3$ of Fig. 1.8 below represent the combined time and shear rate $(\dot{\gamma})$ dependent viscosity loss:

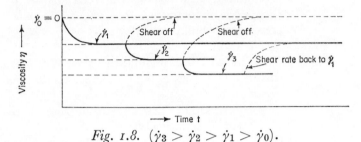

Fig. 1.8. $(\dot{\gamma}_3 > \dot{\gamma}_2 > \dot{\gamma}_1 > \dot{\gamma}_0)$.

If viscosity is plotted vs ln t then it is seen that the viscosity loss is linear with logarithmic time until the equilibrium state is reached. If this is done at a series of arbitrary top shear rates, then a family of parallel straight lines is obtained, Fig. 1.9. Green and Weltmann [3] characterised the slope of these parallel lines $d\eta/d \ln t$ by the symbol B and called it the 'coefficient of thixotropic breakdown with time'. The same workers also established a simple relationship for the 'coefficient of thixotropic breakdown with rate of shear', denoted by the symbol M as defined as the loss in shear stress per unit increase in shear rate [4]. M is in fact—as Green and Weltmann showed— proportional to the area of the hysteresis loop and can be very simply obtained by a technique which we need not go into here but which is fully described in the literature cited.

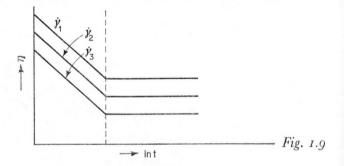

Fig. 1.9

The area of the hysteresis loop is, of course, dependent on the shear acceleration and is not therefore fundamentally significant except in the limiting case of the *equilibrium* hysteresis loop. What *is* fundamentally significant however, is the variation of hysteresis loop area as *a function of* shear acceleration. The *practical* importance of hysteresis loop area determination is realised, however, when industrial processes involving thixotropic (on-machine starch/clay paper coating, lithographic printing, etc.) are carried out within a fairly narrow range of shear acceleration from feed pipe to the region of maximum shear. If this shear acceleration is known, then laboratory tests under simulated conditions will allow reasonable predictions as to the operational behaviour of the thixotropic material to be made. Dahlgren [5] related M and B and showed that one could be calculated from the other. He also analysed Green and Weltmann's system and considered that eight constants are necessary to characterise a thixotropic material fully.

In practical rheology a number of flow curves are obtained or could be thought of which do not appear to conform to any of the types described so far. Some of these can be rationally interpreted whilst others cannot be so interpreted because they arise from experimental artefacts. Curves which may be graphed are shown in Fig. 1.10 below:

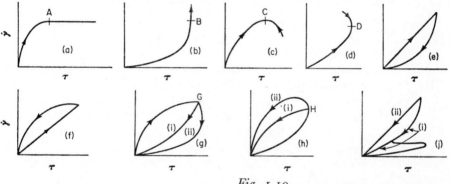

Fig. 1.10

Curve (a), flattening out to a course parallel with the abscissa, is impossible because it implies that the viscosity has reached infinity at and beyond point A—that is to say it has become an ideal Hookean solid. Indeed, even before point A is reached we are dealing with a material with increasing solid state characteristics.

Curve (b) is also impossible, because at point B the curve runs parallel to the ordinate (meaning that the liquid has become inviscid) which is patently absurd. If such a curve is nevertheless occasionally obtained, this is due to 'frictionless' slippage occurring.

Curve (c) is impossible because it is inconceivable that, on increasing the shear stress beyond C, the shear rate should not only drop but that the same path should be traversed on the downcurve.

Curve (d) is impossible for analogous reasons.

Curve (e) is impossible, because it is inconceivable that a Newtonian liquid which shows no demonstrable structural breakdown in the experimental region when making the upcurve should begin to build-up 'structure' on the downward lap.

Curve (f) is equally impossible, because it is inconceivable that a Newtonian liquid which by implication, has no demonstrable shear alignment of 'structure' in the experimental region should show structural breakdown on the downward lap.

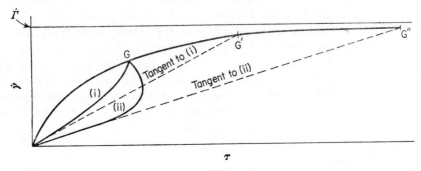

Fig. 1.11

Curves (g) are possible. They have been experimentally observed and we have dealt with them before (Figs. 1.3 and 1.4). The position of point G is a function of the shear acceleration. Provided the instrument can deal with very high shear stresses *and* provided flow still occurs in laminar fashion, it should be possible to shift point G to a position G', or G" where it lies on the tangent through the origin of curves *g* (i) and *g* (ii) respectively—Fig. 1.11 (dotted lines).

In extreme cases (as in curve (ii)), the provisos made are so demanding that the probability of realising this experiment is vanishingly small. The 'liquid' would have preponderantly solid attributes and as the asymptotic shear rate value of $\dot{\Gamma}$ is approached the test material would fracture.

Curve (h) (i) of Fig. 1.10 may just be possible, because one could imagine that the shear acceleration was so high that a memory effect is carried over into the downcurve. There is

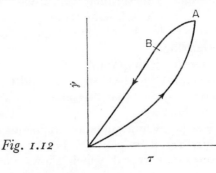

Fig. 1.12

some evidence that such behaviour is encountered occasionally. Some pseudoplastic materials also have a downcurve which curves the other way near the top shear rate (between A and B, Fig. 1.12), before continuing linearly towards the origin. Green and Weltmann had observed this on first up-and-down runs, although they never observed it on subsequent up-and-

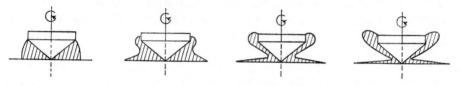

Fig. 1.13. Cavitation in a cone-plate rotational viscometer.

down runs on the same sample, possibly because the material was not allowed to rest sufficiently between runs. They offered no explanation for the phenomenon.

Curve (h) (ii) is sometimes observed in rotational viscometry, but it is spurious. It arises from the fact that under sufficiently high shear rates the specimen will cavitate as it climbs out of the gap—a phenomenon associated with elasticity. As the cavitation effect increases, a situation can arise when the sample is very nearly divided into two portions. Only those regions between the two portions which are still in shearing contact will contribute to the recorded shear stress, which consequently drops substantially. Clearly, we are then obtaining a viscosity which refers partly to the voids which have appeared where material should be found. During cavitation part of the material will flow back into the cavitational void and this produces oscillating shear stresses. The shear geometry is no longer in a steady state and the readings are therefore quite meaningless.

The tendency to cavitate and climb out of the gap is due to 'normal forces' which will be discussed more fully in the next section. Gross cavitation is obvious in a cone/plate viscometer, but spurious effects, even when not visually observed, are always indicated when a curve of type (h) (ii) is recorded. Cup/bob rotational instruments can and do likewise cause cavitation in the material, giving an annular vortex, but spurious curves can be recorded long before the visual manifestation of the phenomenon. It is therefore essential to operate at top shear rates which are well below the critical shear rate for the onset of cavitation. In very viscous materials such as polymer melts this top shear rate is quite low.

The upcurve in (j) is of the kind which is very frequently observed, especially in very viscous liquids. The initial heel portion arises from the fact that the shear acceleration is infinite from the rest position to any small value and only thereafter can one consider the shear acceleration truly constant. We have seen in Fig. 1.5 that the shear acceleration can determine the measured viscosity and it is not surprising, then, that the extremely high shear acceleration which is applied at the commencement of the run produces a correspondingly high measured viscosity. The recoil which follows may either be due to slippage, or to an extremely rapid collapse of the majority of random ground state 'structure', whereupon the liquid may behave as a pseudoplastic with a coincident downcurve (j (i)) or with measurable thixotropy (j (ii)). One could also regard the initial upcurve heel as an inertial effect: It takes a finite time (and this can be a very long time with some liquids) before the applied shear stress waves originating at the boundary communicate themselves through the entire volume of the test specimen and reach a steady resonant state.

It is necessary to point out that thixotropy and rheopexy are isothermal phenomena and that the time dependent viscosity change must be fully reversible to justify those terms. The reverse process may be instantaneous (when no hysteresis loop is experimentally observable) or it may take a very long time indeed.

We have begun to unify the existing concepts of hitherto compartmentalised flow behaviour patterns. We can now come to the following conclusions:

1. All liquids are Newtonian from zero shear rate to some finite shear rate (which may be very low).

2. All liquids are either pseudoplastic or dilatant when they *appear* to start deviating from the Newtonian regime beyond the finite shear rate referred to in (1). They then obey the power law and have a power index of $n \neq 1$.

3. All pseudoplastics reach a condition when such random ground state 'structure' as can be broken down *has* been sheared out. This, again, occurs at some finite shear rate. Beyond this shear rate the liquid enters a region of a *second* Newtonian regime.

4. The Newtonian region preceding a dilatant region is in fact identifiable as the *second* Newtonian region and the apparent 'straight line' which follows the dilatant curvature is not a straight line at all—it is the asymptote of the dilatancy parabola!

5. Liquids which have hitherto been regarded as Newtonian only happen to be such that the finite shear rate necessary to make the power index depart from unity is so high as to be experimentally unattainable. Alternatively, the initial Newtonian, the power law and the second Newtonian regions may

all be coincident because the power index in the power-law region happens to be unity, but the probability of this happening is vanishingly small.

6. All pseudoplastics show an anticlockwise hysteresis loop when an upcurve and downcurve is run ($\dot{\gamma}$ as ordinate) in which the area of the hysteresis loop may vary from zero to the value corresponding to a linear downcurve returning to the origin. All pseudoplastics are therefore time dependent (although the time dependency may happen to be so small as to escape experimental observation) and all pseudoplastics may thus be regarded as additionally thixotropic by definition.

7. All dilatants, i.e. liquids in which the power index $n > 1$, show a clockwise hysteresis loop when an upcurve and a downcurve is run ($\dot{\gamma}$ as ordinate) in which the area of the hysteresis loop may vary from zero to values which may be in excess of that corresponding to a linear downcurve returning to the origin, although a linear downcurve should theoretically be obtained if the upcurve is run at equilibrium shear acceleration. All dilatants are therefore time dependent (although the time dependency may happen to be so small as to escape experimental observation) and thus show rheopexy by definition.

In order to complete the unification [6] we also have to account for the 'plastic' and Bingham body behaviour and show how they can be fitted into this seven-point system. This is not difficult.

A Bingham body can be said to have an extremely high initial viscosity, so that it appears to be in the solid state and has an initial upcurve which is coincident with the shear stress axis. If this represents the initial Newtonian region which ends at the yield value, then we have the second Newtonian region following on directly, the intervening pseudoplastic region is vanishingly small in extent and cannot be demonstrated experimentally, but that does not necessarily mean that it is non-existent. A 'plastic' would then simply be a Bingham body in which the intervening pseudoplastic region *can* be experimentally demonstrated.

We should also be able to account for a flow behaviour of the type first described by Ostwald in 1925 (Fig. 1.14). In this curve we are supposed to have a first Newtonian region up to A, a pseudoplastic region from A to B, a dilatant region from B to C and another Newtonian region beyond C.

In general, this curve fits into the scheme in that we have an initial and second Newtonian region and an intermediate non-Newtonian region. But within the non-Newtonian region we also have an inversion from pseudoplastic to dilatant. This could well occur for instance, in a disperse system in which 'structure' is first broken down, but in which the continuous phase is insufficient to lubricate the disperse phase adequately as the shear rate is increased beyond a certain point or in which

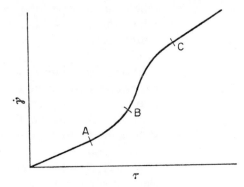

Fig. 1.14.
Ostwald flow curve.

vortex structures appear. Indeed, one may argue that this is the most general state of liquids, if one interpolates another linear portion into the curve at the point of inversion (Fig. 1.15).

Point B in Ostwald's flow curve has now been developed into a linear stretch between B_1 and B_2 which would represent the second Newtonian region. At extremely high shear rates we cannot, of course, expect laminar flow to persist and because of the onset of turbulence the dilatant region from B_2 to C will not always be readily observed. However—see the references to the work of Van der Vegt and Smit [1] and of Galt and Maxwell [2], and on pp. 11–12 of this chapter. The curve beyond C will be asymptotic to a limiting shear rate, it is in fact a continuation of the curve from B_2 and is *not* truly linear. It does *not*, therefore, represent a third Newtonian region.

The concept of the generalised flow curve makes it necessary to add an eighth postulate to the scheme:

8. All liquids show dilatant behaviour beyond the second Newtonian region, provided the nature of the material makes it possible to maintain non-turbulent flow under the corresponding shear rate conditions. This leads to a generalised flow curve from which all other basic types can be derived, viz:

(i) *Newtonian:* A material with generalised flow curve in which experimental conditions beyond point A are not realisable.

(ii) *Pseudoplastics:* A material with generalised flow curve in which experimental conditions can be realised which enable one to reach some point between A and B_2 whilst maintaining laminar flow.

(iii) *Dilatant*: A material with generalised flow curve in which the first Newtonian, the pseudoplastic and possibly part or all of the second Newtonian region are traversed from zero shear rate to such a small shear rate that these regions become telescoped and are inseparable by experimental observation.

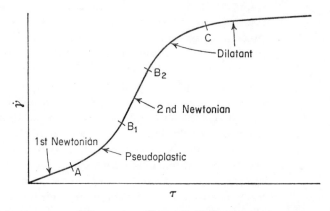

Fig. 1.15.   Generalised flow curve.

(iv) *'Plastic'*: A material with generalised flow curve in which the first Newtonian region is virtually coincident with the shear stress axis and which is followed by the pseudoplastic and possibly part of the second Newtonian region.

(v) *'Bingham body'*: A material with generalised flow curve in which the first Newtonian region is virtually coincident with the shear stress axis and in which the pseudoplastic region is vanishingly small, so that the first Newtonian region appears to be immediately followed by the second Newtonian region. The dilatant part may not be readily experimentally realisable.

(vi) *Ostwald flow*: This characterises a material with generalised flow curve in which the second Newtonian region has degenerated into a point of inversion which marks an apparently direct transition from the pseudoplastic to the dilatant condition.

The reader is strongly recommended to bear the concept of the generalised flow curve well in mind, because we shall need to return to it in Chapter 8 when it will be seen to constitute a powerful tool in the consolidation of liquid and solid states into a unified fluid state of multiple aspect.

3. *Liquids with significantly large instantaneous elastic deformation*

So far we have ignored the manifestations of Hookean behaviour in liquids on the assumption that these are negligibly small, but in liquids of high viscosity this assumption is patently unjustified. The highly viscous polymer melts show substantial instantaneous elasticity as evidenced by die swelling of extrudate and by an increase in thickness of calendered material as it leaves the calender nip.

In the introduction reference has already been made to the importance of the experimental time scale in relation to the type of response of a material to an applied stress. In the preceding section (2, p. 14) this time dependence showed itself in the creation of 'structural' variations characteristic for any given shear stress which in turn gives a dynamic equilibrium of the build-up and breakdown processes. The ratio of these processes represents a fundamental constant, provided sufficient time has elapsed to allow equilibrium to become established. In this situation the sheared liquid is able to adjust itself *internally*, that is to say, to absorb energy reversibly within itself by the creation or destruction of transient spatial configurations of its constituent volume elements. Such a situation is common and visible in disperse systems (flocculation and deflocculation). In as much as the new spatial arrangement differs energetically from that of the random ground state to which it will revert in time on removal of the stress, it represents stored energy, i.e. elasticity. But the recovery of this stored energy is *not instantaneous*—it is retarded by the internal viscosity of the material. Its characteristic parameter is the time constant of reversibility known as the *retardation time*.

The type of elastic response which we shall be considering now differs from retarded elasticity in that the elastic response is not exclusively an internal arrangement: the rearrangement additionally transmits energy to its environment. This energy *is* instantaneously recovered on removal of the stress. Indeed this becomes the only elastic response available to the liquid when the time of applied stress is so short that the processes of flow and/or of retarded elasticity cannot get under way. There must be an upper limit of stress application beyond which the material will begin to relax and instantaneous elasticity will become less prominent. This time limit is related to the *relaxation time*.

Retardation and relaxation are two aspects of viscoelasticity to which we will return in later chapters.

Instantaneous elasticity can be easily recognised in the laboratory. Weissenberg has first accounted for the well known tendency of some viscous liquids to climb up a stirrer on reaching a critical shear rate. The effect can also be observed when a cone with vertical channels and with its axis normal to a plate is rotated with a viscous liquid placed into the gap between cone and plate. There is a tendency for the liquid to climb into those channels, as can be seen in Fig. 1.16 below.

The arrows in Fig. 1.16 indicate that a force exists in a direction perpendicular to the shear plane under those conditions. This force can be measured without drilling channels into the cone (and thus losing material from the gap) by placing

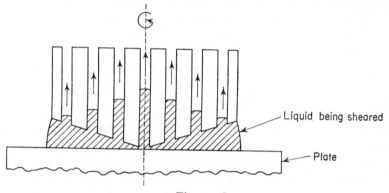

Liquid being sheared

Plate

Fig. 1.16

pressure transducers in contact with the liquid at the cone interface and by measuring the pressure ('normal force') exerted by the liquid which tends to lift the cone from its point of contact with the plate. This is the basis of 'rheogoniometry', the study of instantaneous elasticity or of tensile forces in liquids when they are subjected to nominal shear conditions.

It is easy to see why this situation arises *par excellence* in polymer melts. We have here a system of long chains in random configuration which are entangled amongst themselves. In the first Newtonian region true viscous flow can occur because the chain segments are freely movable and because there is sufficient time (low shear rate!) for the loose entanglements to unmake and remake themselves without affecting the random ground state structure. In the pseudoplastic region (higher shear rate) the ground state structure will transform itself in accordance with Le Chatelier's principle, in such a way that the units of flow will become increasingly aligned in the direction of flow. If the reattainment of the ground state takes a relatively long time, then we have thixotropy or retarded elasticity. But at even higher shear rates the entanglements will have no chance to unravel, consequently they will tend to tighten up frictionally. The reaction to the energy supplied then exceeds the additive response of:

1. dissipation through flow, and

2. storage in the form of reversible non-ground-state configurations.

The *third* response which must therefore be added to make up the total response is a further distinct storage response: Distortion of chemical bond angles and elongation of bond distances. This will exert a hydrostatic pressure on the containing surfaces of the liquid which can then be measured as a 'normal force'. The order of magnitude of these forces is much greater than that which is involved in mere 'structure' changes, and so long as no actual melt fracture occurs, the energy involved is instantaneously recovered on removal of the stress.

In order that instantaneous elasticity should assert itself it is necessary for entanglement of this kind to be possible and the pre-condition for entanglement is the existence of a chain structure of sufficient length. In non-polymeric systems and disperse systems chains can also occur but they will have 'bonds' created by Van der Waals and London forces which are much weaker than those involved in the primary chemical bonds of polymer chains. Instantaneous elasticity will therefore be much less pronounced in disperse systems than in polymer melts. Low molecular weight polymers are somewhat intermediate in this respect; they show little instantaneous elasticity. Ordinary polymers in which the entanglements have been 'combed out' by filtration in solution (a procedure resorted to in the molecular weight fractionation of polymers), also show much less instantaneous elasticity than the original polymer when the desolvated fractions are quantitatively reunited and converted to melt. Indeed the instantaneous elasticity is only gradually restored after considerable periods of maintaining the recombined fractions in the melt condition which proves that the gradual reentanglement of combed out fractions is directly responsible for instantaneous elasticity of polymer melts. It is not even necessary to separate the fractions in the first place—solution, filtration and speedy solvent flash-off is quite sufficient to reduce the instantaneous elasticity of the melt to a remarkable degree. We shall return to this again in Chapter 4 when considering the causes of melt elasticity.

From the very nature of polymer melts it is inevitable that entanglements should normally exist in abundance and that the melts should also be highly viscous, except at unusually high temperatures. It is therefore inevitable that normal forces of a high order should develop during shear even at low shear rates. This makes it impossible to realise shear rates in rotational viscometry which would be high enough for the 'post-pseudoplastic' features of the generalised flow curve to manifest themselves; even if they did appear, they would be masked by the effects of the normal forces. In order to minimise normal forces and in order to demonstrate as much as possible of the generalised flow curve one would have to work with dilute polymer solutions rather than melts. This has been done by Brodnyan, Gaskins and Phillippoff [7] who demonstrated the existence of a second Newtonian region in decalin solutions of polyisobutylene. Their data were subsequently used by Wright and Crouse [8] in their development of a new concept of generalising the flow data of pseudoplastics.

The power law—the foundation of rheological measurement of fluids

Let us consider the power law as previously stated:

$$\tau = \eta \dot{\gamma}^n$$

Taking logarithms we get:

$$\log \tau = \log \eta + n \log \dot{\gamma}$$

It follows that in a log/log plot we should obtain a straight line with slope n. The intercept $\log \eta$ cannot be evaluated because the extrapolation to $\dot{\gamma} = 0$ ($\log \dot{\gamma} = -\infty$) can obviously not be carried out. Instead it is customary to extrapolate to $\dot{\gamma} = 1 \sec^{-1}$ ($\log \dot{\gamma} = 0$) and to use the viscosity at unit shear rate as a reference point.

From the generalised flow curve we know that the power index n is not constant over the entire idealised shear rate range. But the first Newtonian region in polymer melts is exceedingly degenerate and the subsequent pseudoplastic region is characterised by a constant value of n which is less than unity. This is invariably found in laminar shear of polymer melts and this flow regime persists:

(a) Until the second Newtonian region is reached in which case n again becomes unity, *or*

(b) In the event of the second Newtonian region being degenerate, until the dilatant region is reached, in which case n becomes a constant value *in excess of* unity, *or*

(c) Until normal forces become so large that the effects associated with purely viscous flow and 'structure' changes become submerged, *or*

(d) Until flow disturbances supervene and cause melt fracture.

In polymer processing a change in flow regime is usually associated with the transitions from a pseudoplastic to a dilatant flow regime and the abrupt changes in the slope of the log/log plot has received attention from numerous workers [9–13].

Since this change in flow regime must eventually result in melt fracture and so produce gross distortion in an extrudate, the shear stress (as governed by the pressure applied) and the resulting production rate has an upper limit which it is clearly vital for the trade extruder and the machine designer to be aware of. A study of the log/log plot at a number of possible processing temperatures will supply this information at a glance.

We can therefore regard the power index as a constant value at any one melt temperature within the useful processing range of shear rates and shear stresses.

The power law has been recognised and used for many years and it has been accepted as an empirical working tool which, within its limitations, gives satisfactory results. The limitations, of course, consist of the fact that it cannot cope with the additional complications which arise from instantaneous elasticity and which manifest themselves in die swell. It is remarkable that until recently, no theoretical explanation for the existence of the power law has been forthcoming. It is even more remarkable that when Scott Blair [14] put forward his theoretic explanation this explanation proved to be exceedingly simple:

Consider a liquid with much random ground 'structure' which, on shearing at increasing shear rate (but at constant shear acceleration) attains an equilibrium stress τ for each value of $\dot{\gamma}$. Assume that we are dealing with a pseudoplastic liquid of zero thixotropy and that the number of linkages (S) which contribute to the random ground 'structure' is linearly related to $\log \tau$. This assumption which is the simplest one possible cannot hold at infinitely low shear stress, since $dS/d\tau$ would be $-\infty$ at $\tau = 0$.

But once the structure has begun to break down (i.e. when the initial Newtonian region has been traversed), one might reasonably suppose that the rate of breakdown will decrease exponentially as the number of linkages is reduced, so that

$$\tau = k\, e^{-aS}$$

where a and k are constants, and

$$\ln \tau = k' - aS$$

This equation may be written:

Equation 1.12
$$-\frac{dS}{d\tau} = \frac{a}{\tau}$$

When we maintain or even increase the shear rate $\dot{\gamma}$, however, we are preventing the elements which make up the 'structure' from remaining in contact long enough for the restoring forces to reestablish the 'structure'—that is to say, to increase S.

Again, it would be simplest to assume that the number of linkages which can remain intact despite the applied shear rate $\dot{\gamma}$ will be a linear function of log $\dot{\gamma}$. This leads to an equation analogous to (1.12):

Equation 1.13
$$-\frac{dS}{d\dot{\gamma}} = \frac{b}{\dot{\gamma}}$$

where b is another constant.

This means that S, the number of linkages per unit volume, is proportional to the logs of both τ and $\dot{\gamma}$.

Dividing (1.12) by (1.13) we get

$$\frac{d\dot{\gamma}}{d\tau} = \frac{a}{b}\frac{\dot{\gamma}}{\tau}$$

or

$$\frac{d\dot{\gamma}}{\dot{\gamma}} = \frac{a}{b}\frac{d\tau}{\tau}$$

and on integration:

Equation 1.14
$$\ln \dot{\gamma} = \frac{a}{b}\ln \tau + C$$

where C is the integration constant. This is identical with the commonly used empirical equation of the power law in logarithmic form.

Equation 1.14 can, of course, hold only over a limited range of shear rates. At very high shear rates all linkages may be broken and there would then be a straight line portion in the $\dot{\gamma}$ vs τ curve, so that a second Newtonian region must appear.

The case of dilatancy can also be accommodated in this argument. Here we get an increase in 'structure' with increasing shear rate because the frequency and duration of contact between the 'structural' elements is increasing. This causes an increase in the number of linkages and the only difference would be that equations 1.12 and 1.13 would have a positive sign. Equation 1.14 would be unaffected.

The significance of constants a and b is that of rate constants for the breakdown and build-up processes respectively:

In a Newtonian a and b are identical,

In a pseudoplastic $a > b$,

In a dilatant $b > a$.

Writing Equation 1.3 in the non-logarithmic form, it becomes

Equation 1.15
$$\dot{\gamma} = C\tau^{a/b} \quad \text{or} \quad \tau = \text{const.}\ \dot{\gamma}^{b/a}$$

from which it is obvious that the ratio of the rate constants b/a is identical with the power index n.

We have come to the conclusion that the development of the power-law characteristics where the power index $n \neq 1$ and the appearance of time dependence of

(*a*) thixotropy/rheopexy, and

(*b*) conventional viscoelasticity,

are all due to 'structure' changes in the liquid under shear. We have briefly considered how the forces arise which are the causes of such 'structure'. In polymers these forces are of a very high magnitude since they involve primary chemical bond distances and bond angles and since additional secondary valence forces such as hydrogen bonding may also be involved. Rather weaker forces also play a part and these are variously referred to as London forces, interfacial forces and electrostatic forces. These weaker forces may be relatively unimportant in polymer melts in the face of the much more powerful forces mentioned previously, but in disperse systems these weaker forces may be the only ones present and so assume a decisive importance.

There is another system which would repay the trouble of rheological investigation, namely the rheological behaviour of certain types of bacteria suspended in a suitable medium. Some bacteria tend to form linear chains of organism (*streptococci*), others form grape-like clusters (*staphylococci*). The forces which link these assemblages together are probably of appreciable magnitude, although they are certainly not primary valence forces. The bacterial suspensions, however, have the following attributes which make them experimentally attractive as model materials:

1. Streptococcal chains are formally similar to polymer chains in that they can become entangled once they have reached a certain critical length and once they have reached this length it will be segments rather than the chain as a whole which constitutes the unit of flow when a suspension is agitated.

2. The chains and their 'structural' rearrangement in a rotational shear field could be directly observed or recorded by high speed ciné photography under a microscope, provided the organisms are stained *in vivo* and they may thus be of considerable significance to an observer who is a plastics rheologist.

3. Bacteria are of an order of magnitude intermediate between, say, pigments in suspension and the molecular repeating units of a polymer chain. Although this order of magnitude of disperse phase can be achieved in emulsion polymerisation,

the resulting disperse polymer particles do not form chains and grape clusters as bacteria do.

4. The effect of chain structures could be assessed by using staphylococcal clusters as controls in comparison with streptococcal chains.

5. By careful investigation of the variables it may be possible to demonstrate the different regions of the generalised flow curve.

6. By using various concentrations of bacterial suspensions the effect of crowding may be conveniently investigated and compared to, say, PVC pastes.

7. Graphing of shear stress vs time at constant shear rate and otherwise constant conditions would make it possible to obtain a continuous automatic record of bacterial reproduction rates.

8. It would be possible to examine the effect of changes in the disperse phase from a spherical to a cylindrical shape by switching from bacteria to bacilli and also to investigate other geometries such as those presented by some yeasts. Branched structures such as those developed by dendritic penicillium moulds could be investigated, and, again, useful inferences or pointers may be obtained for polymer rheology.

All this may be some way from our principal field of activity. But evidence obtained from whatever testing ground can contribute to a better understanding of the laws governing the properties of matter and so can make possible further advances in diverse fields which otherwise have few, if any, direct connections. The power law is only one of a number of equations put forward on empirical or theoretical grounds by various people [15] and these will not be considered here. Some comment, however, is appropriate on an equation suggested recently by Cross [16] in a paper given at the Joint Conference on the Advances in polymer Science and Technology held in London in September 1966 and jointly sponsored by the Plastics and Polymer Group of the Society of Chemical Industry, the Institution of the Rubber Industry and the Plastics Institute. The equation, based on the assumption that the viscosity of a non-Newtonian liquid is a continuous function of the shear rate (as indeed is the power law) takes the form:

$$\eta = \eta_\infty + \frac{\eta_0 - \eta_\infty}{1 + \alpha \dot{\gamma}^n}$$

where η is the viscosity at shear rate $\dot{\gamma}$ and η_0 and η_∞ 'are limiting values of the viscosity at respectively zero and infinite rate of shear and α is a constant.' Cross found that this equation was widely applicable to solutions, suspensions, pastes and melts and that a value of $\frac{2}{3}$ for n provides a remarkably good fit with experimental results. The equation obviously covers the range of the generalised flow curve from zero shear rate to the second Newtonian region and Cross's η_∞ should therefore be regarded as the second Newtonian viscosity, rather than a genuine 'infinite shear rate' viscosity. If the viscosity is plotted against log $\dot{\gamma}$, then a sigmoid curve is obtained (Fig. 1.17a).

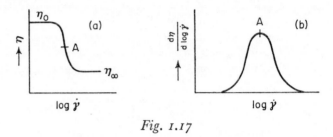

Fig. 1.17

If the first derivative $d\eta/d \log \dot{\gamma}$ is plotted vs log $\dot{\gamma}$, then a maximum is obtained at the point of inversion characterised by a shear rate which Cross denotes as $\dot{\gamma}_c$ (Fig. 1.17b). The peak denotes the midpoint of a Gaussian transition region from first to second Newtonian behaviour and is therefore descriptive of the liquid in its pseudoplastic range.

If η in Cross's equation is set equal to η_c (the viscosity corresponding to a shear rate of $\dot{\gamma}_c$), then η_c is the arithmetic mean of η_0 and η_∞. It also follows that $\gamma_c = \alpha^{-1/n}$. The most interesting aspect of this lies in the fact that this treatment: (*a*) has its exact analogy in the modulus/log deformation and modulus/time (frequency) relationships which apply to the solid state and which will be dealt with in subsequent chapters, that it is (*b*) independent of the type of deformation involved, and (*c*) that it applies to the entire spectrum of fluid materials, irrespective of whether they are nominally classed as 'solids' or 'liquids'.

2 Vectors and tensors. Fundamental equations

Definitions. Reasons for using tensors. Vector and tensor notation: Subscripts, the 'nabla function', dot and cross products, divergence, gradient, curl, multiple dot products in Cartesian co-ordinates. Polar co-ordinates. Components in Cartesian, cylindrical and spherical co-ordinates (conductive heat flux vector, stress tensor). The equations of continuity, momentum and energy.

VECTORS are entities which possess direction as well as magnitude. Thus, a force applied to one point of a body is a *vector* since it is characterised by both magnitude and direction. If the force is applied not merely at one point but at many points simultaneously with interacting effects, as for example in a continuous body subjected to tension, then we are dealing with a vector of a higher order which we call a *tensor*. Conversely, we could regard vectors as degenerate tensors.

If we chose any arbitrary system of reference co-ordinates in three dimensions such that the point of application of the force coincides with the origin, then we can represent a vector by a straight line headed by an arrow indicative of its direction and by a length indicative of its magnitude. Velocity is a vector since it obviously has both magnitude and direction.

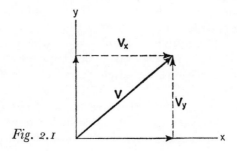

Fig. 2.1

If we represent the velocity vector in a plane by $\mathbf{v}$ (Fig. 2.1) we can clearly resolve it algebraically by referring to its components $\mathbf{v}_x$ and $\mathbf{v}_y$. In other words, we can realise a vector by applying component forces $\mathbf{v}_x$ and $\mathbf{v}_y$ to a point in such a way that a resultant force $\mathbf{v}$ is obtained. This was known to prehistoric man who used bows and slings; he would apply two component forces to the missile and would propel the missile: (1) with a velocity which was a function of the magnitude of these forces, and (2) in a direction which was a function of the direction of the two component forces. These functions are additive, so that one can write:

$$\mathbf{v} = \mathbf{v}_x + \mathbf{v}_y$$

but the rules of addition are different from those which apply to directionless magnitudes or *scalars*.

We are not concerned here with the *derivation* of the rules which apply to the manipulation of vectors. Suffice it to say that all algebraic operations (addition and subtraction, multiplication and division, integration and differentiation) can be carried out in an appropriate manner the justification of which can be found in standard mathematical textbooks.

It is, however, necessary to state that the scalar and directional attributes of a vector can be separated for reasons of algebraic convenience and that this is done by introducing the concept of the *unit vectors* **i, j, k,** such that:

$$\mathbf{v}_x = \mathbf{i}v_x$$
$$\mathbf{v}_y = \mathbf{j}v_y$$
$$\mathbf{v}_z = \mathbf{k}v_z$$

where **i, j, k** always have a magnitude of unity and a direction coincident with that of the appropriate reference co-ordinate and where v_x, v_y, v_z are scalars.

It will be apparent that with different algebraic operational methods it is essential to use a method of notation which differs from that which is used in the manipulation of scalars. There are additional compelling reasons for such a notation, namely:

1. The number of components which have to be considered and which make up our operating equations may be large and the expressions, although generally of a simple cyclic nature, become cumbersome. Vector notation provides a shorthand way of expressing them clearly and enables one to carry out the necessary manipulations in a simple manner. The final expression can then be expanded using one's memory or tables.

2. The final equations are rigorous and make no simplifying assumptions. If assumptions become necessary in order to obtain a numerical solution from experimental data which may be incomplete because of difficulties in determining certain components, then these assumptions and corresponding provisos in the assessment of the result can be made. This leaves the

door open for refining the result by more rigorous treatment in the future as experimental techniques become more sophisticated.

3. In addition to being concise, equations in vector notation are independent of arbitrary co-ordinate systems. The expansion of the final equation will take the form appropriate to the co-ordinate system which is defined by the geometry of the force field, but it is not necessary to pay any attention to the latter during the generalised algebraic manipulation until the final equation is obtained.

Vector and tensor notation

It is necessary to define the notation which will be used in the following. It is not the *only* notation in common use, but it is a reasonably simple one.

(i) Vectors and tensors are characterised in print by bold type. In writing it is convenient to denote them by a squiggle under the symbol.

(ii) Single subscripts denote a component. Thus, a vector $\mathbf{v}$ has a component v_x with unit vector $\mathbf{i}$ and magnitude v_x. The velocity vector in a rectangular coordinate system x, y, z, will therefore be completely defined by the unit vectors $\mathbf{i}, \mathbf{j}, \mathbf{k}$ and the component scalars v_x, v_y, v_z.

(iii) If more than one subscript is present, then we are dealing with the components of a tensor, characterised by the following convention: The second subscript defines the direction of the force and the first subscript defines the plane which is perpendicular to the stated axis. For example, in a system of

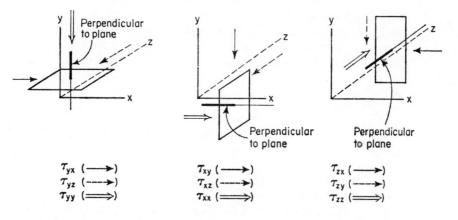

Fig. 2.2. The components of the stress tensor $\boldsymbol{\tau}$.

rectangular co-ordinates, the stress tensor τ will have the components shown in Fig. 2.2:

It is seen that the stress tensor has nine components. Those with equal subscripts represent normal (tensile) forces, whilst those with unequal subscripts represent tangential (shear) forces. It is therefore clear that the shear stress tensor can be represented by an array, the full development of which is:

$$
\begin{array}{ccc}
\tau_{xx} & \tau_{xy} & \tau_{xz} \\
\tau_{yx} & \tau_{yy} & \tau_{yz} \\
\tau_{zx} & \tau_{zy} & \tau_{zz}
\end{array}
$$

(iv) The *nabla function* or 'vector operator'. ∇ is a shorthand method of writing the expression:

$$
\mathbf{i}\,\frac{\partial}{\partial x} + \mathbf{j}\,\frac{\partial}{\partial y} + \mathbf{k}\,\frac{\partial}{\partial z}
$$

for rectangular co-ordinates.

It never stands alone, since the partial differentials must state the parameter which is subjected to the operation.

Thus, when we write $\nabla\phi$, we mean that parameter ϕ is subjected to the operation:

$$
\left(\mathbf{i}\,\frac{\partial}{\partial x} + \mathbf{j}\,\frac{\partial}{\partial y} + \mathbf{k}\,\frac{\partial}{\partial z}\right)\phi = \mathbf{i}\,\frac{\partial\phi}{\partial x} + \mathbf{j}\,\frac{\partial\phi}{\partial y} + \mathbf{k}\,\frac{\partial\phi}{\partial z}
$$

for rectangular co-ordinates.

$\nabla\phi$ is also called the 'gradient of ϕ' or 'grad ϕ', where ϕ is a scalar.

(v) Vector multiplication, like vector addition, has its special algebraic rules. These need not be derived here; but it must be stated that there exist two kinds of products, a '*scalar product*' or '*dot product*' and a '*vector product*' or '*cross product*'.

(vi) The *scalar or dot product* is written $\mathbf{p}\,.\,\mathbf{q}$. If it involves the nabla function (which is itself a vector), then $\nabla\,.\,\mathbf{a}$ is called the '*divergence of a*' or '*div a*', so that

$$
\operatorname{div}\mathbf{a} = \nabla\,.\,\mathbf{a} = \left(\mathbf{i}\,\frac{\partial}{\partial x} + \mathbf{j}\,\frac{\partial}{\partial y} + \mathbf{k}\,\frac{\partial}{\partial z}\right)\left(\mathbf{i}a_x + \mathbf{j}a_y + \mathbf{k}a_z\right)
$$

div $\mathbf{a}$ therefore involves the scalar entities of the components a_x, a_y and a_z of the vector $\mathbf{a}$.

It must be added that by the rules of vector manipulation the products of the unit vectors $\mathbf{i}$, $\mathbf{j}$ and $\mathbf{k}$ are as follows:

$$
\left.
\begin{array}{ll}
\mathbf{i} \cdot \mathbf{i} = 1 & \mathbf{i} \cdot \mathbf{j} = 0 \\
\mathbf{j} \cdot \mathbf{j} = 1 & \mathbf{i} \cdot \mathbf{k} = 0 \\
\mathbf{k} \cdot \mathbf{k} = 1 & \mathbf{j} \cdot \mathbf{k} = 0 \\
& \mathbf{j} \cdot \mathbf{i} = 0 \\
& \mathbf{k} \cdot \mathbf{i} = 0 \\
& \mathbf{k} \cdot \mathbf{j} = 0
\end{array}
\right\}
or:
\begin{array}{ll}
\mathbf{i} \cdot \mathbf{j} = 1 & \text{when} \quad \mathbf{i} = \mathbf{j} \\
\mathbf{i} \cdot \mathbf{j} = 0 & \text{when} \quad \mathbf{i} \neq \mathbf{j}
\end{array}
$$

so that

$$
\text{div } \mathbf{a} = \mathbf{\nabla} \cdot \mathbf{a} = \frac{\partial a_x}{\partial x} + \frac{\partial a_y}{\partial y} + \frac{\partial a_z}{\partial z}
$$

The quantities on the right-hand side of this expression are the scalar entities of the components of the vector **a** and their sum is therefore a scalar.

In general, the dot product of two vectors is a scalar and this is the reason why the dot product is also called (paradoxically at first sight, but nevertheless logically) the scalar product.

(viii) The '*vector product*' or '*cross product*' of two vectors is written **p** ∧ **q**. If it involves the nabla function (which is itself a vector), then $\mathbf{\nabla} \wedge \mathbf{b}$ is called the 'curl of *b*' or '*curl b*', where

$$
\text{curl } b = \mathbf{\nabla} \wedge \mathbf{b} = \left(\mathbf{i} \frac{\partial}{\partial x} + \mathbf{j} \frac{\partial}{\partial y} + \mathbf{k} \frac{\partial}{\partial z} \right) \wedge \left(\mathbf{i} b_x + \mathbf{j} b_y + \mathbf{k} b_z \right)
$$

According to the rules of cross product multiplication, the order of writing the multiplicands (unlike in ordinary algebra) is *not* interchangeable, but

$$
\mathbf{p} \wedge \mathbf{q} = -\mathbf{q} \wedge \mathbf{p}
$$

Furthermore, the cross products of unit vectors are as follows:

$$
\begin{array}{ll}
\mathbf{i} \wedge \mathbf{i} = 0 & \mathbf{i} \wedge \mathbf{j} = -\mathbf{j} \wedge \mathbf{i} = \mathbf{k} \\
\mathbf{j} \wedge \mathbf{j} = 0 & \mathbf{j} \wedge \mathbf{k} = -\mathbf{k} \wedge \mathbf{j} = \mathbf{i} \\
\mathbf{k} \wedge \mathbf{k} = 0 & \mathbf{k} \wedge \mathbf{i} = -\mathbf{i} \wedge \mathbf{k} = \mathbf{j}
\end{array}
$$

Now, if the right-hand side of the curl **b** equation is multiplied out we obtain:

$$
\text{curl } b = \mathbf{\nabla} \wedge \mathbf{b} = \mathbf{i} \wedge \mathbf{i} \frac{\partial b_x}{\partial x} + \mathbf{i} \wedge \mathbf{j} \frac{\partial b_y}{\partial x} + \mathbf{i} \wedge \mathbf{k} \frac{\partial b_z}{\partial x} +
$$

$$
+ \mathbf{j} \wedge \mathbf{i} \frac{\partial b_x}{\partial y} + \mathbf{j} \wedge \mathbf{j} \frac{\partial b_y}{\partial y} + \mathbf{j} \wedge \mathbf{k} \frac{\partial b_z}{\partial y} +
$$

$$
+ \mathbf{k} \wedge \mathbf{i} \frac{\partial b_x}{\partial z} + \mathbf{k} \wedge \mathbf{j} \frac{\partial b_y}{\partial z} + \mathbf{k} \wedge \mathbf{k} \frac{\partial b_z}{\partial z}
$$

39

5

Making use of the stated cross products of unit vectors, the terms involving $\mathbf{i} \wedge \mathbf{i}$, $\mathbf{j} \wedge \mathbf{j}$ and $\mathbf{k} \wedge \mathbf{k}$ disappear. Substituting the third unit vector (with the appropriate sign) for the cross product of each pair of unit vectors as it appears, we obtain:

$$\text{curl } \mathbf{b} = \nabla \wedge \mathbf{b} = \mathbf{k} \frac{\partial b_y}{\partial x} - \mathbf{j} \frac{\partial b_z}{\partial x} -$$

$$- \mathbf{k} \frac{\partial b_x}{\partial y} + \mathbf{i} \frac{\partial b_z}{\partial y} +$$

$$+ \mathbf{j} \frac{\partial b_x}{\partial z} - \mathbf{i} \frac{\partial b_y}{\partial z}$$

or

$$\text{curl } \mathbf{b} = \nabla \wedge \mathbf{b} = \mathbf{i} \left(\frac{\partial b_z}{\partial y} - \frac{\partial b_y}{\partial y} \right) +$$

$$+ \mathbf{j} \left(\frac{\partial b_x}{\partial z} - \frac{\partial b_z}{\partial x} \right) +$$

$$+ \mathbf{k} \left(\frac{\partial b_y}{\partial x} - \frac{\partial b_x}{\partial y} \right)$$

A strict cyclic arrangement is clearly in evidence. It will be noted that the right-hand side of the above equation is identical with the expansion of the determinant:

$$\begin{vmatrix} \mathbf{i} & \mathbf{j} & \mathbf{k} \\ \dfrac{\partial}{\partial x} & \dfrac{\partial}{\partial y} & \dfrac{\partial}{\partial z} \\ b_x & b_y & b_z \end{vmatrix}$$

It will also be noted that whilst the dot product of two vectors is a scalar, the cross product of two vectors is a vector the components of which are the three right-hand side terms of the equation for the curl.

(viii) Another useful shorthand expression is that of the *multiple-dot product* which is used to indicate that the summation is to be applied to more than one kind of component. This does not arise with vectors (tensors of the first order) which have only components of one kind, but it *does* arise with tensors of the second order because they have components of two kinds, namely those

(*a*) where the subscripts are identical, and those,

(*b*) where the subscripts are non-identical.

If
$$\nabla \cdot \boldsymbol{\theta} = \frac{\partial}{\partial x} \theta_x + \frac{\partial}{\partial y} \theta_y + \frac{\partial}{\partial z} \theta_z$$

then

$$\mathbf{v}(\mathbf{\nabla\theta}) = v_x \frac{\partial \theta_x}{\partial x} + v_y \frac{\partial \theta_y}{\partial y} + v_z \frac{\partial \theta_z}{\partial z}$$

where $\mathbf{v}$ is a vector (say, velocity).

If, on the other hand, we treat the second order tensor $\mathbf{\tau}$ in an analogous fashion, then we have two types of components, namely the normal force (tensile) components of the kind τ_{ii} and the tangential force (shear) components of the kind τ_{ij}, where i and j (in rectangular co-ordinates) can have any of the connotations x, y and z in turn.

In such a case, in order to emphasise that there are two kinds of forces involved which have to be summed, it is convenient to write, *not* $\mathbf{\tau} . (\mathbf{\nabla v})$, but $\mathbf{\tau} : (\mathbf{\nabla v})$.

Using the same rules as before, the expansion of $\mathbf{\tau} : (\mathbf{\nabla} . \mathbf{v})$ becomes (for rectangular co-ordinates) in general:

$$\mathbf{\tau} : (\mathbf{\nabla v}) = \Sigma_{ii} \tau_{ii}(\mathbf{\nabla v}) + \Sigma_{\substack{i \\ j}} \tau_{\substack{i \\ j}}(\mathbf{\nabla v})$$

or in particular:

$$\mathbf{\tau} : (\mathbf{\nabla} . \mathbf{v}) = \left\{ \tau_{xx} \frac{\partial v_x}{\partial x} + \tau_{yy} \frac{\partial v_y}{\partial y} + \tau_{zz} \frac{\partial v_z}{\partial z} \right\} +$$

$$+ \left\{ (\tau_{xy} + \tau_{yx}) \left(\frac{\partial v_x}{\partial y} + \frac{\partial v_y}{\partial x} \right) + (\tau_{xz} + \tau_{zx}) \left(\frac{\partial v_x}{\partial z} + \frac{\partial v_z}{\partial x} \right) + \right.$$

$$\left. + (\tau_{yz} + \tau_{zy}) \left(\frac{\partial v_y}{\partial z} + \frac{\partial v_z}{\partial y} \right) \right\}$$

Whilst the expansion can be easily and purely mechanically written down without being an undue burden on the memory, it is obviously extremely cumbersome and quite unnecessary if the elegant and perfectly unambiguous left-hand expression can be used instead.

Another fact emerges from the expansion however: If normal force components are so small as to be negligible compared to the shear force components, then the first three terms can be ignored. Furthermore, if $\tau_{ij} \equiv \tau_{ji}$ (as is the case in symmetrical tensors) then the expansion can be simplified to:

$$\frac{\mathbf{\tau} : (\mathbf{\nabla} . \mathbf{v})}{2} = \left\{ \tau_{xy} \left(\frac{\partial v_x}{\partial y} + \frac{\partial v_y}{\partial x} \right) + \tau_{yz} \left(\frac{\partial v_y}{\partial z} + \frac{\partial v_z}{\partial y} \right) + \right.$$

$$\left. + \tau_{zx} \left(\frac{\partial v_z}{\partial x} + \frac{\partial v_x}{\partial z} \right) \right\}$$

A double dot product is used in conjunction with the strain rate tensor $\mathbf{\Delta}$ with which we shall be dealing in due course.

Other co-ordinate systems

So far it has been assumed that we are only dealing with rectangular co-ordinates. But it is obvious that other co-ordinate systems are equally capable of defining the position of points in space.

Of the co-ordinate systems other than the rectangular those of the greatest practical importance are the cylindrical and the spherical systems.

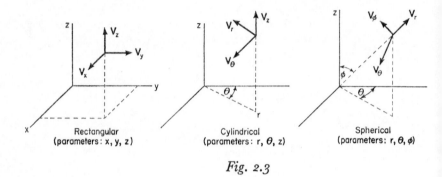

Rectangular
(parameters: x, y, z)

Cylindrical
(parameters: r, θ, z)

Spherical
(parameters: r, θ, φ)

Fig. 2.3

Assuming that a velocity vector **v** acts on a point in space, then this vector can be resolved into component vectors in the following ways (Fig. 2.3):

In order to convert the expansions of grad **v**, div **v** and curl **v** from the rectangular to the cylindrical or spherical co-ordinate system, it is only necessary to substitute the correct function of r and θ for x and y in the cylindrical system, and the correct functions of r, θ and ϕ for x, y and z in the spherical system. In extruder flow problems the cylindrical geometry is most important.

The way in which a vector can be split into its components in rectangular, cylindrical and spherical co-ordinates is perhaps most usefully illustrated in the case of the *heat flux vector*: Heat itself is a scalar, but the conductive transfer of heat from point A to point B involves a gradient along the line AB. This means that conductive heat flux is directional and therefore a vector.

The scalar entities of the components of the conductive heat flux vector **q** are as follows:

TABLE I*

Rectangular	Cylindrical	Spherical
$q_x = -k \dfrac{\partial T}{\partial x}$	$q_r = -k \dfrac{\partial T}{\partial r}$	$q_r = -k \dfrac{\partial T}{\partial r}$
$q_y = -k \dfrac{\partial T}{\partial y}$	$q_\theta = -k \dfrac{1}{r} \dfrac{\partial T}{\partial \theta}$	$q_\theta = -k \dfrac{1}{r} \dfrac{\partial T}{\partial \theta}$
$q_z = -k \dfrac{\partial T}{\partial z}$	$q_z = -k \dfrac{\partial T}{\partial z}$	$q_\phi = -k \dfrac{1}{r \sin \theta} \dfrac{\partial T}{\partial \phi}$

where k is the thermal conductivity of the material in question.

In extrusion flow problems it is, of course, necessary to express the stress tensor in the terms in which the data are stated. The velocity in the axial direction of an extruder (the z-axis in a cylindrical geometry) is generally available. There is also flow in the radial direction; this is possibly accessible with considerable experimental sophistication and may not be negligibly small. Flow in the third direction should also be included in a rigorous flow equation, but this is not experimentally determinable so far and is, in any case, thought to be negligibly small.

Resolving the stress tensor **τ** into its components the scalar entities of which are τ_{ii} and τ_{ij} (and of which, as has been shown, there are a total of nine) is not sufficient. In order to solve flow problems, the stress tensor **τ** must ultimately be expressed in terms of the scalars of the velocity components which are relevant to the prevailing geometry. Tables for conversion of the stress tensor components have been worked out and are reproduced below in slightly modified form with the permission of the authors and publishers [1].

In these expansions η_N is the Newtonian viscosity of the liquid and K the compressibility.

* Reprinted with permission from R. B. Bird, W. E. Steward, and E. N. Lightfoot, *Transport Phenomena*, John Wiley (1960).

TABLE II. Components of the Stress Tensor τ

1. Rectangular Co-ordinates

(a)
$$\begin{cases} \tau_{xx} = \eta_N \left[2 \frac{\partial v_x}{\partial x} - \frac{2}{3} (\nabla \cdot \mathbf{v}) \right] + K(\nabla \cdot \mathbf{v}) \\[2mm] \tau_{yy} = \eta_N \left[2 \frac{\partial v_y}{\partial y} - \frac{2}{3} (\nabla \cdot \mathbf{v}) \right] + K(\nabla \cdot \mathbf{v}) \\[2mm] \tau_{zz} = \eta_N \left[2 \frac{\partial v_z}{\partial z} - \frac{2}{3} (\nabla \cdot \mathbf{v}) \right] + K(\nabla \cdot \mathbf{v}) \end{cases}$$

where

$$(\nabla \cdot \mathbf{v}) = \operatorname{div} \mathbf{v} = \frac{\partial v_x}{\partial x} + \frac{\partial v_y}{\partial y} + \frac{\partial v_z}{\partial z}$$

(b)
$$\begin{cases} \tau_{xy} = \tau_{yx} = \eta_N \left[\frac{\partial v_x}{\partial y} + \frac{\partial v_y}{\partial x} \right] \\[2mm] \tau_{yz} = \tau_{zy} = \eta_N \left[\frac{\partial v_y}{\partial z} + \frac{\partial v_z}{\partial y} \right] \\[2mm] \tau_{zx} = \tau_{xz} = \eta_N \left[\frac{\partial v_z}{\partial x} + \frac{\partial v_x}{\partial z} \right] \end{cases}$$

2. Cylindrical Co-ordinates

(a)
$$\begin{cases} \tau_{rr} = \eta_N \left[2 \frac{v_r}{r} - \frac{2}{3} \operatorname{div} \mathbf{v} \right] + K \operatorname{div} \mathbf{v} \\[2mm] \tau_{\theta\theta} = \eta_N \left[2 \left(\frac{1}{r} \frac{\partial v_\theta}{\partial \theta} + \frac{v_r}{r} \right) - \frac{2}{3} \operatorname{div} \mathbf{v} \right] + K \operatorname{div} \mathbf{v} \\[2mm] \tau_{zz} = \eta_N \left[2 \frac{\partial v_z}{\partial z} - \frac{2}{3} \operatorname{div} \mathbf{v} \right] + K \operatorname{div} \mathbf{v} \end{cases}$$

where

$$\operatorname{div} \mathbf{v} = \frac{1}{r} \frac{\partial}{\partial r} r v_r + \frac{1}{r} \frac{\partial v_\theta}{\partial \theta} + \frac{\partial v_z}{\partial z}$$

(b)
$$\begin{cases} \tau_{r\theta} = \tau_{\theta r} = \eta_N \left[r \frac{\partial}{\partial r} \left(\frac{v_\theta}{r} \right) + \frac{1}{r} \frac{\partial v_r}{\partial \theta} \right] \\[2mm] \tau_{\theta z} = \tau_{z\theta} = \eta_N \left[\frac{\partial v_\theta}{\partial z} + \frac{1}{r} \frac{\partial v_z}{\partial \theta} \right] \\[2mm] \tau_{zr} = \tau_{rz} = \eta_N \left[\frac{\partial v_z}{\partial r} + \frac{\partial v_r}{\partial z} \right] \end{cases}$$

3. Spherical Co-ordinates

(a)
$$\begin{cases} \tau_{rr} = \eta_N \left[2 \frac{\partial v_r}{\partial r} - \frac{2}{3} \operatorname{div} \mathbf{v} \right] + K \operatorname{div} \mathbf{v} \\[2mm] \tau_{\theta\theta} = \eta_N \left[2 \left(\frac{1}{r} \frac{\partial v_\theta}{\partial \theta} + \frac{v_r}{r} \right) - \frac{2}{3} \operatorname{div} \mathbf{v} \right] + K \operatorname{div} \mathbf{v} \\[2mm] \tau_{\phi\phi} = \eta_N \left[2 \left(\frac{1}{r \sin \theta} \frac{\partial v_\phi}{\partial \phi} + \frac{v_r}{r} + \frac{v_\theta \cot \theta}{r} \right) - \frac{2}{3} \operatorname{div} \mathbf{v} \right] + K \operatorname{div} \mathbf{v} \end{cases}$$

TABLE II *Concluded*

(b)

$$\begin{cases} \tau_{r\theta} = \tau_{\theta r} = \eta_N \left[r \frac{\partial}{\partial r} \left(\frac{v_\theta}{r} \right) + \frac{1}{r} \frac{\partial v_r}{\partial \theta} \right] \\[2mm] \tau_{\theta\phi} = \tau_{\phi\theta} = \eta_N \left[\frac{\sin \theta}{r} \frac{\partial}{\partial \theta} \left(\frac{v_\phi}{\sin \theta} \right) + \frac{1}{r \sin \theta} \frac{\partial v_\theta}{\partial \phi} \right] \\[2mm] \tau_{\phi r} = \tau_{r\phi} = \eta_N \left[\frac{1}{r \sin \theta} \frac{\partial v_r}{\partial \phi} + r \frac{\partial}{\partial r} \left(\frac{v_\phi}{r} \right) \right] \end{cases}$$

where

$$\operatorname{div} \mathbf{v} = \frac{1}{r^2} \frac{\partial}{\partial r} (r^2 v_r) + \frac{1}{r \sin \theta} \frac{\partial}{\partial \theta} (v_\theta \sin \theta) + \frac{1}{r \sin \theta} \frac{\partial v_\phi}{\partial \phi}$$

The equations in Table II look formidably complex, but it should be noted that they can be generalised (by the shorthand notation described earlier on) as follows:

(a) $$\tau_{ii} = \eta_N \left[2 \frac{\partial v_i}{\partial i} - \frac{2}{3} \operatorname{div} \mathbf{v} \right] + K \operatorname{div} \mathbf{v}$$

(b) $$\tau_{ij} = \tau_{ji} = \eta_N \left[\frac{\partial v_i}{\partial j} + \frac{\partial v_j}{\partial i} \right]$$

where i and j can have any value of $\begin{Bmatrix} x, y, z \\ r, \theta, z \\ r, \theta, \phi \end{Bmatrix}$ in a $\begin{Bmatrix} \text{rectangular} \\ \text{cylindrical} \\ \text{spherical} \end{Bmatrix}$

co-ordinate system, as the case may be.

They are further simplified in practice as follows: If the liquid in flow is assumed to be incompressible, $K = 0$ then, despite the fact that large pressure drops occur along an extruder channel, the velocity components v_i are constant in the i-direction. Therefore div $\mathbf{v}$ (see expansion, p. 38) and the first term in the bracket of equations (a) disappear. This means that the normal forces can be ignored. If the further approximation is made that in an extruder and die channel the velocity vector components

$$\frac{\partial}{\partial r} \left(\frac{v_\theta}{r} \right), \frac{\partial v_r}{\partial \theta}, \frac{\partial v_\theta}{\partial z}, \frac{\partial v_z}{\partial \theta} \quad \text{and} \quad \frac{\partial v_r}{\partial z}$$

are all zero, then the only velocity vector component remaining is $\partial v_z / \partial r$ and the stress tensor is represented by the magnitude of τ_{rz} alone and all other components disappear, so that:

the *magnitude* of $\boldsymbol{\tau}$ is given by $\eta_N (\partial v_z/\partial r)$, a result which it is extremely simple to apply and which we shall be using extensively in due course.

In aerodynamics where the fluid is a gas and not a liquid, compressibility is, of course a most important factor even under isothermal conditions and div $\mathbf{v}$ cannot therefore be ignored. Fortunately we do not need to concern ourselves with the great complexities which arise as a consequence since we are exclusively interested in polymer melts.

The equations of continuity, momentum and energy [2]

These equations formulate fundamental principles of physics. They are not in themselves sufficient to evaluate flow problems since they are independent of the rheological behaviour of the liquid, but together with the appropriate rheological equation they provide all the necessary data for solving flow problems.

A volume element of liquid in motion is defined by the scalar quantities of density (ρ), pressure (P), temperature (T) and by the velocity vector $\mathbf{v}$. The stress $\boldsymbol{\tau}$ which results from the shear forces is a tensor of the second order. The continuity equation considers these parameters as a function of time (t) and position as defined by the components in any generalised system of co-ordinates.

Imagine that the volume element is isolated from the bulk of the moving volume but continues to follow the motion of the latter without exerting any force upon it. The volume element can then exchange energy—but not matter—with the bulk volume across its boundaries. Since, according to the principle of the conservation of matter, the mass of a closed system remains constant, any local increase or decrease in density with time within the volume element must be exactly reflected by the rate of flow of material into or away from such loci, so that the overall density change within the volume element is zero. This is expressed by the *continuity equation*

$$\frac{\mathrm{d}\rho}{\mathrm{d}t} = -\rho(\boldsymbol{\nabla} \cdot \mathbf{v}) = -\rho \operatorname{div} \mathbf{v}$$

On expanding this general equation for rectangular, cylindrical and spherical co-ordinates the following equations are obtained:

Rectangular co-ordinates (x, y, z) :

$$\frac{\mathrm{d}\rho}{\partial t} = -\left[\frac{\partial(\rho v_x)}{\partial x} + \frac{\partial(\rho v_x)}{\partial y} + \frac{\partial(\rho v_z)}{\partial z}\right]$$

Cylindrical co-ordinates (r, θ, z) :

$$\frac{\partial\rho}{\partial t} = -\left[\frac{\partial(\rho_r v_r)}{r\,\partial r} + \frac{\partial(\rho v_\theta)}{r\,\partial\theta} + \frac{\partial(\rho v_z)}{\partial z}\right]$$

Spherical co-ordinates (r, θ, ϕ) :

$$\frac{\partial\rho}{\partial t} = -\left[\frac{\partial(\rho_r^2 v_r)}{r^2\,\partial r} + \frac{\partial(\rho v_\theta \sin\theta)}{r\sin\theta\,\partial\theta} + \frac{\partial(\rho v_\phi)}{r\sin\theta\,\partial\phi}\right]$$

If the liquid can be assumed to be incompressible, then div **v** will, of course, again be zero and no local changes can occur in the density of the volume element as a function of time. This is the case in polymer melts during processing, but obviously, if the enclosed material is gaseous this assumption does not hold and tremendous complications may be expected to arise, for example, in the extrusion of expanded polystyrene. The continuity equation was included here for the sake of completeness and because it demonstrates, once again, the elegant and convenient way in which a relationship can be generally expressed through the use of vector and tensor notation.

The momentum equation

Newton's second law of motion requires [2] 'that the rate of increase of momentum of the fluid element be equal to the sum of all the forces acting upon it.'

The rate of increase of momentum is equal to $\rho\,\mathrm{d}\mathbf{v}/\mathrm{d}t$.

The body forces are represented by a function of:

(i) the pressure differential in the nett direction of flow; this is a negative quantity given by $-\nabla P$ or grad P,

(ii) the stress tensor; this is given by $\nabla \cdot \boldsymbol{\tau}$ or div $\boldsymbol{\tau}$,

(iii) the gravitational force; this is given by $\rho\mathbf{g}$, where the vector **g** represents 'the resultant of the body forces (per unit mass) acting on the fluid at any point' [2].

The equation is therefore generally expressed as follows:

$$\rho\frac{\mathrm{d}\mathbf{v}}{\mathrm{d}t} = -\nabla P + \nabla \cdot \boldsymbol{\tau} + \rho\mathbf{g} = -\operatorname{grad} P + \operatorname{div} \boldsymbol{\tau} + \rho\mathbf{g}$$

and can therefore be expanded as in Table III [1]:

TABLE III

Rectangular Co-ordinates

x-component:

$$\rho\left(\frac{\partial v_x}{\partial t} + v_x\frac{\partial v_x}{\partial x} + v_y\frac{\partial v_x}{\partial y} + v_z\frac{\partial v_x}{\partial z}\right) = -\frac{\partial P}{\partial x} + \frac{\partial \tau_{xx}}{\partial x} + \left(\frac{\partial \tau_{yx}}{\partial y} + \frac{\partial \tau_{zx}}{\partial z}\right) + \rho g_x$$

-component:

$$\rho\left(\frac{\partial v_y}{\partial t} + v_x\frac{\partial v_y}{\partial x} + v_y\frac{\partial v_y}{\partial y} + v_z\frac{\partial v_y}{\partial z}\right) = -\frac{\partial P}{\partial y} + \left(\frac{\partial \tau_{xy}}{\partial x} + \frac{\partial \tau_{yy}}{\partial y} + \frac{\partial \tau_{zy}}{\partial z}\right) + \rho g_y$$

z-component:

$$\rho\left(\frac{\partial v_z}{\partial t} + v_x\frac{\partial v_z}{\partial x} + v_y\frac{\partial v_z}{\partial y} + v_z\frac{\partial v_z}{\partial z}\right) = -\frac{\partial P}{\partial z} + \left(\frac{\partial \tau_{xz}}{\partial x} + \frac{\partial \tau_{yz}}{\partial y} + \frac{\partial \tau_{zz}}{\partial z}\right) + \rho g_z$$

Cylindrical Co-ordinates

r-component:

$$\rho\left(\frac{\partial v_r}{\partial t} + v_r\frac{\partial v_r}{\partial r} + \frac{v_\theta}{r}\frac{\partial v_r}{\partial \theta} - \frac{v_\theta^2}{r} + v_z\frac{\partial v_r}{\partial z}\right)$$
$$= -\frac{\partial P}{\partial r} + \left[\frac{1}{r}\frac{\partial}{\partial r}(r\tau_{rr}) + \frac{1}{r}\frac{\partial \tau_{r\theta}}{\partial \theta} - \frac{\tau_{\theta\theta}}{r} + \frac{\partial \tau_{rz}}{\partial z}\right] + \rho g_r$$

θ-component:

$$\rho\left(\frac{\partial v_\theta}{\partial t} + v_r\frac{\partial v_\theta}{\partial r} + \frac{v_\theta}{r}\frac{\partial v_\theta}{\partial \theta} + \frac{v_r v_\theta}{r} + v_z\frac{\partial v_\theta}{\partial z}\right)$$
$$= -\frac{1}{r}\frac{\partial P}{\partial \theta} + \left[\frac{1}{r^2}\frac{\partial}{\partial r}(r^2\tau_{r\theta}) + \frac{1}{r}\frac{\partial \tau_{\theta\theta}}{\partial \theta} + \frac{\partial \tau_{\theta z}}{\partial z}\right] + \rho g_\theta$$

z-component:

$$\rho\left(\frac{\partial v_z}{\partial t} + v_r\frac{\partial v_z}{\partial r} + \frac{v_\theta}{r}\frac{\partial v_z}{\partial \theta} + v_z\frac{\partial v_z}{\partial z}\right) = -\frac{\partial P}{\partial z} + \left[\frac{1}{r}\frac{\partial}{\partial r}(r\tau_{yz}) + \frac{1}{r}\frac{\partial \tau_{\theta z}}{\partial \theta} + \frac{\partial \tau_{zz}}{\partial z}\right] + \rho g_z$$

Spherical Co-ordinates

r-component:

$$\rho\left(\frac{\partial v_r}{\partial t} + v_r\frac{\partial v_r}{\partial r} + \frac{v_\theta}{r}\frac{\partial v_r}{\partial \theta} + \frac{v_\phi}{r\sin\theta}\frac{\partial v_r}{\partial \phi} - \frac{v_\theta^2 + v_\phi^2}{r}\right)$$
$$= -\frac{\partial P}{\partial r} + \left[\frac{1}{r^2}\frac{\partial}{\partial n}(r^2\tau_{rr}) + \frac{1}{r\sin\theta}\frac{\partial}{\partial \theta}(\tau_{r\theta}\sin\theta) + \frac{1}{r\sin\theta}\frac{\partial \tau_{r\phi}}{\partial \phi} - \frac{\tau_{\theta\theta} + \tau_{\phi\phi}}{r}\right] + \rho g_r$$

θ-component:

$$\rho\left(\frac{\partial v_\theta}{\partial t} + v_r\frac{\partial v_\theta}{\partial r} + \frac{v_\theta}{r}\frac{\partial v_\theta}{\partial \theta} + \frac{v_\phi}{r\sin\theta}\frac{\partial v_\theta}{\partial \phi} + \frac{v_r v_\theta}{r} - \frac{v_\phi^2\cot\theta}{r}\right)$$
$$= -\frac{1}{r}\frac{\partial P}{\partial \theta} + \left[\frac{1}{r^2}\frac{\partial}{\partial r}(r^2\tau_{r\theta}) + \frac{1}{r\sin\theta}\frac{\partial}{\partial \theta}(\tau_{\theta\theta}\sin\theta) + \frac{1}{r\sin\theta}\frac{\partial \tau_{\theta\phi}}{\partial \phi} + \frac{\tau_{r\theta}}{r} - \frac{\cot\theta}{r}\tau_{\phi\phi}\right] + \rho g_\theta$$

ϕ-component:

$$\rho\left(\frac{\partial v_\phi}{\partial t} + v_r\frac{\partial v_\phi}{\partial r} + \frac{v_\theta}{r}\frac{\partial v_\phi}{\partial \theta} + \frac{v_\phi}{r\sin\theta}\frac{\partial v_\phi}{\partial \phi} + \frac{v_\phi v_r}{r} + \frac{v_\theta v_\phi}{r}\cot\theta\right)$$
$$= -\frac{1}{r\sin\theta}\frac{\partial P}{\partial \phi} + \left[\frac{1}{r^2}\frac{\partial}{\partial r}(r^2\tau_{r\phi}) + \frac{1}{r}\frac{\partial \tau_{\theta\phi}}{\partial \theta} + \frac{1}{r\sin\theta}\frac{\partial \tau_{\phi\phi}}{\partial \phi} + \frac{\tau_{r\phi}}{r} + \frac{2\cot\theta}{r}\tau_{\theta\phi}\right] + \rho g_\phi$$

In polymer melts ρg is negligibly small compared to the other forces acting upon the liquid. We can also assume incompressibility, so that div τ disappears. Lastly, one can make the simplifying assumption that most of the velocity vector components in the cylindrical geometry of an extruder are zero. The result, then, is suitably simple and can be used to obtain solutions to flow problems.

The Energy Equation is based on the principle of conservation of energy as embodied in the first law of thermodynamics. In tensor notation it takes the form:

$$\rho C_v \frac{\mathrm{d}T}{v\,\mathrm{d}T} = - [\nabla \cdot \mathbf{q}] - T\left(\frac{\partial P}{\partial T}\right)_\rho [\nabla \cdot \mathbf{v}] + [\tau : \nabla \cdot \mathbf{v}]$$

TABLE IV

Rectangular Co-ordinates

$$\rho C_v \left(\frac{\partial T}{\partial t} + v_x \frac{\partial T}{\partial x} + v_y \frac{\partial T}{\partial y} + v_z \frac{\partial T}{\partial z}\right) = - \left[\frac{\partial q_x}{\partial x} + \frac{\partial q_y}{\partial y} + \frac{\partial q_z}{\partial z}\right] -$$
$$- T\left(\frac{\partial P}{\partial T}\right)_\rho \left(\frac{\partial v_x}{\partial x} + \frac{\partial v_y}{\partial y} + \frac{\partial v_z}{\partial z}\right) + \left\{\tau_{xx}\frac{\partial v_x}{\partial x} + \tau_{yy}\frac{\partial v_y}{\partial y} + \tau_{zz}\frac{\partial v_z}{\partial z}\right\} +$$
$$+ \left\{\tau_{xy}\left(\frac{\partial v_x}{\partial y} + \frac{\partial v_y}{\partial x}\right) + \tau_{xz}\left(\frac{\partial v_x}{\partial z} + \frac{\partial v_z}{\partial x}\right) + \tau_{yz}\left(\frac{\partial v_y}{\partial z} + \frac{\partial v_z}{\partial y}\right)\right\}$$

Cylindrical Co-ordinates

$$\rho C_v \left(\frac{\partial T}{\partial t} + v_r \frac{\partial T}{\partial r} + \frac{v_\theta}{r}\frac{\partial T}{\partial \theta} + v_z \frac{\partial T}{\partial z}\right) = - \left[\frac{1}{r}\frac{\partial}{\partial r}(rq_r) + \frac{1}{r}\frac{\partial q\theta}{\partial \theta} + \frac{\partial q_z}{\partial z}\right] -$$
$$- T\left(\frac{\partial P}{\partial T}\right)_p \left(\frac{1}{r}\frac{\partial}{\partial r}(rv_r) + \frac{1}{r}\frac{\partial v_\theta}{\partial \theta} + \frac{\partial v_z}{\partial z}\right) + \left\{\tau_{rr}\frac{\partial v_r}{\partial r} + \tau_{\theta\theta}\frac{1}{r}\left(\frac{\partial v_\theta}{\partial \theta} + v_r\right) + \tau_{zz}\frac{\partial v_z}{\partial z}\right\} +$$
$$+ \left\{\tau_{r\theta}\left[r\frac{\partial}{\partial r}\left(\frac{v_\theta}{r}\right) + \frac{1}{r}\frac{\partial v_r}{\partial \theta}\right] + \tau_{rz}\left(\frac{\partial v_z}{\partial r} + \frac{\partial v_r}{\partial z}\right) + \tau_{\theta z}\left(\frac{1}{r}\frac{\partial v_z}{\partial \theta} + \frac{\partial v_\theta}{\partial z}\right)\right\}$$

Spherical Co-ordinates

$$\rho C_v \left(\frac{\partial T}{\partial t} + v_r \frac{\partial T}{\partial r} + \frac{v_\theta}{r}\frac{\partial T}{\partial \theta} + \frac{v_\phi}{r\sin\theta}\frac{\partial T}{\partial \phi}\right) =$$
$$= - \left[\frac{1}{r^3}\frac{\partial}{\partial r}(r^2 q_r) + \frac{1}{r\sin\theta}\frac{\partial}{\partial \theta}(q_\theta \sin\theta) + \frac{1}{r\sin\theta}\frac{\partial q_\phi}{\partial \phi}\right] -$$
$$- T\left(\frac{\partial P}{\partial T}\right)_\rho \left(\frac{1}{r^2}\frac{\partial}{\partial r}(r^2 v_r) + \frac{1}{r\sin\theta}\frac{\partial}{\partial \theta}(v_\theta \sin\theta) + \frac{1}{r\sin\theta}\frac{\partial v_\phi}{\partial \phi}\right) +$$
$$+ \left\{\tau_{rr}\frac{\partial v_r}{\partial r} + \tau_{\theta\theta}\left(\frac{1}{r}\frac{\partial v_\theta}{\partial \theta} + \frac{V_r}{r}\right) + \tau_{\phi\phi}\left(\frac{1}{r\sin\theta}\frac{\partial v_\phi}{\partial \phi} + \frac{v_r}{r} + \frac{v_\theta \cot\theta}{r}\right)\right\} +$$
$$+ \left\{\tau_{r\theta}\left(\frac{\partial v_\theta}{\partial r} + \frac{1}{r}\frac{\partial v_\theta}{\partial \theta} - \frac{v_\theta}{r}\right) + \tau_{r\phi}\left(\frac{\partial v_\phi}{\partial r} + \frac{1}{r\sin\theta}\frac{\partial v_r}{\partial \phi} - \frac{v_\phi}{r}\right) +$$
$$+ \tau_{\theta\phi}\left(\frac{1}{r}\frac{\partial v_\phi}{\partial \theta} + \frac{1}{r\sin\theta}\frac{\partial v_\theta}{\partial \phi} - \frac{\cot\theta}{r}v_\phi\right)\right\}$$

where C_v is the specific heat at constant volume and $\mathbf{q}$, the conductive heat flux vector is given by Fourier's law of heat conduction:

$$\mathbf{q} = -k\nabla T$$

where k is the thermal conductivity of the material.

In compressible fluids such as gases, aerosols and smokes $[\nabla . \mathbf{v}]$ or div $\mathbf{v}$ is clearly of the greatest importance, but in polymer melts it is zero if incompressibility is assumed, so that the energy equation again reduces to a very simply manageable expression. The expansion of the conductive heat flux vector has already been given (p. 43).

The full rigorous expansion of the tensor form of the energy equation is given in Table IV above which is again taken from Bird, Steward and Lightfoot [1] and which again illustrates by its many terms and its consequent cumbersome nature what enormous advantages accrue from the use of generalised tensor notation.

An Equation of State is required for problems in which the density is variable. This is manifestly important for gases, aerosols, smokes, and for materials in the temperature regions where discontinuous states coexist. But it is of no importance in the typical processing of polymer melts.

Equations which describe the Temperature and Pressure Dependence of Other Fluid Properties may be required. An example of this is the well known Arrhenius equation which applies to rate processes. This will be considered in Chapter 4.

3 The rheological equation for the liquid state. Uses of the equations

The general rheological equation, its power-law form and transformations thereof. The strain-rate tensor and its expression in various co-ordinate systems. Use of the fundamental equations and the rheological equation for the determination of velocity and temperature profiles in flow: Cartesian and cylindrical geometries with Newtonian and general power-law liquids under known boundary conditions.

A NUMBER of rheological equations appear in the literature, but we shall confine ourselves to a consideration of the power-law equation and its special case which applies to Newtonian behaviour, because polymer melt flow problems can be readily solved if the power law applies, as it does under laminar flow conditions.

In the most general form the rheological equation is simply expressed by

$$\boldsymbol{\tau} = f(\dot{\gamma})$$

As the power law, it takes the form $\boldsymbol{\tau} = \eta\dot{\gamma}^n$ which reduces to the Newtonian case in the special event of the power index n being equal to unity.

The power law [1] can be transformed and so rendered even more useful by the following procedure:

Let τ^0 and $\dot{\gamma}^0$ be the shear stress and shear rate respectively at some arbitrary standard reference state, say, when $\dot{\gamma}^0 = 1$ sec^{-1}, or when $\tau^0 = 1$ dyne cm^{-2}. If, in general $\boldsymbol{\tau} = \eta\dot{\gamma}^n$, then $\tau^0 = \eta^0(\dot{\gamma}^0)^n$, and

$$\frac{\tau}{\tau^0} = \frac{\eta}{\eta^0}\left(\frac{\dot{\gamma}}{\gamma^0}\right)^n$$

whence it can be seen that

$$\eta = \eta^0\left(\frac{\dot{\gamma}}{\dot{\gamma}^0}\right)^{n-1} \quad \text{or} \quad \eta = \eta^0\left(\frac{\tau}{\tau^0}\right)^{(n-1)/n}$$

The strain rate tensor

Having previously indicated the nature of the stress tensor τ, it is easy to see that strain rate $\dot\gamma$ is also a tensor function, for it has both magnitude and direction with respect to any chosen set of space co-ordinates. To emphasise this it is convenient to denote it with the symbol Δ and to write the general rheological equation for a Newtonian: $\tau = \eta_N \Delta$ and for a power-law fluid: $\tau = \eta\Delta$, incompressibility ($\nabla \cdot \mathbf{v} = 0$) being assumed.

The strain rate tensor, then, has components defined by:

$$\Delta_{ij} = \left(\frac{\partial v_i}{\partial x_j}\right) + \left(\frac{\partial v_j}{\partial x_i}\right)$$

Now, η_N (the Newtonian viscosity) is, by definition, independent of the strain (shear rate), but η (the viscosity for a power-law fluid) *is* dependent on $\dot\gamma$ ($\eta = f(\Delta)$). Viscosity is a scalar quantity and can therefore only be a function of the scalar *invariants* of Δ, that is to say, of those scalar quantities that remain unchanged upon rotation of the co-ordinate axes. There are three such invariants (e.g. in rectangular co-ordinates, $\Delta_{xy}, \Delta_{xz}, \Delta_{yz}$), two of which are zero in simple shear. In simple shear the only nonzero velocity component is v_x and the only nonzero derivative of v_x is $(\partial v_x/\partial y)$—(ref: parallel plate flow diagram Fig. 3.1). If, in a power-law fluid, $\tau = \eta\Delta$ and $\eta = f(\Delta)$ then this double dependence of τ on Δ is indicated, by tensor notation convention, by a *double* dot product $\Delta:\Delta$.*

The generalised form of the power law can then be written:

$$\eta = \eta^0 \left[\frac{(\Delta:\Delta)}{2}\right]^{(n-1)/2}$$

where the standard state (superscript o) is taken at a shear rate of 1 sec^{-1}. The dot product of vectors is a scalar, as is the double dot product of tensors. The full expansion of the scalar double dot product ($\Delta:\Delta$) is given in the following table for the usual three co-ordinate types, but having already observed that in simple shear these rigorous expansions have mostly zero terms, then the expression, e.g. for rectangular co-ordinates, reduces to:

$$\frac{\Delta:\Delta}{2} = \left(\frac{\partial v_x}{\partial y}\right)^2$$

* $(\Delta:\Delta) = \Sigma_i\Sigma_j(\Delta_{ij})^2$.

and equation

$$\eta = \eta^0 \left(\frac{\dot{\gamma}}{\dot{\gamma}^0}\right)^{n-1}$$

becomes (for $\dot{\gamma}^0 = 1$):

$$\eta = \eta^0 \left(\frac{\partial v_x}{\partial y}\right)^{n-1}$$

whence

$$\eta = \eta^0 \left[\frac{(\Delta:\Delta)}{2}\right]^{(n-1)/2}$$

The strain rate tensor Δ differs from the shear rate $\dot{\gamma}$ in the following:

The strain rate tensor Δ is a three-dimensional entity and therefore provides a rigorous solution to a flow problem if its components are known. The shear rate $\dot{\gamma}$ represents the unidimensional degeneration of the strain rate tensor which is obtained when two of the three shear components of the strain rate tensor as well as all the normal force components are set to zero. This cannot provide rigorous solutions to flow problems, but it does have the advantage of simplicity at the sacrifice of accuracy.

TABLE V. The Function $\frac{1}{2}(\Delta:\Delta)$ Expressed in the Usual Co-ordinate Systems [2]

___

(a) Rectangular

$$\frac{1}{2}(\Delta:\Delta) = 2\left[\left(\frac{\partial v_x}{\partial x}\right)^2 + \left(\frac{\partial v_y}{\partial y}\right)^2 + \left(\frac{\partial v_z}{\partial z}\right)^2\right] +$$

$$+ \left[\frac{\partial v_y}{\partial x} + \frac{\partial v_x}{\partial y}\right]^2 + \left[\frac{\partial v_z}{\partial y} + \frac{\partial v_y}{\partial z}\right]^2 + \left[\frac{\partial v_x}{\partial z} + \frac{\partial v_z}{\partial x}\right]^2$$

(b) Cylindrical

$$\frac{1}{2}(\Delta:\Delta) = 2\left[\left(\frac{\partial v_r}{\partial r}\right)^2 + \left(\frac{1}{r}\frac{\partial v_\theta}{\partial \theta} + \frac{v_r}{r}\right)^2 + \left(\frac{\partial v_z}{\partial z}\right)^2\right] +$$

$$+ \left[r\frac{\partial}{\partial r}\left(\frac{v_\theta}{r}\right) + \frac{1}{r}\frac{\partial v_r}{\partial \theta}\right]^2 + \left[\frac{1}{r}\frac{\partial v_z}{\partial \theta} + \frac{\partial v_\theta}{\partial z}\right]^2 + \left[\frac{\partial v_n}{\partial z} + \frac{\partial v_z}{\partial r}\right]^2$$

(c) Spherical

$$\frac{1}{2}(\Delta:\Delta) = 2\left[\left(\frac{\partial v_r}{\partial r}\right)^2 + \left(\frac{1}{r}\frac{\partial v_\theta}{\partial \theta} + \frac{v_r}{r}\right)^2 + \left(\frac{1}{r\sin\theta}\frac{\partial v_\phi}{\partial \phi} + \frac{v_r}{r} + \frac{v_\theta\cot\theta}{r}\right)^2\right] +$$

$$+ \left[r\frac{\partial}{\partial r}\left(\frac{v_\theta}{r}\right) + \frac{1}{r}\frac{\partial v_r}{\partial \theta}\right]^2 + \left[\frac{\sin\theta}{r}\frac{\partial}{\partial \theta}\left(\frac{v_\phi}{\sin\theta}\right) + \frac{1}{r\sin\theta}\frac{\partial v_\theta}{\partial \phi}\right]^2 +$$

$$+ \left[\frac{1}{r\sin\theta}\frac{\partial v_r}{\partial \phi} + r\frac{\partial}{\partial r}\left(\frac{v_\phi}{r}\right)\right]^2$$

___

We have seen that the power law can be expressed in terms involving the strain rate tensor Δ and this is the best form of using it. The expansion of the function $(\Delta:\Delta/2)$ which this involves is given in the usual co-ordinate systems in Table V. It will be obvious that these expansions can be greatly simplified if certain components or certain derivatives of these components can be set to zero. The ultimate simplification leads to the recovery of the unidimensional power law and the reappearance of the unidimensional shear rate $\dot{\gamma}$.

Viscosity equations other than the power law can be generalised into tensor notation (and so made independent of specific co-ordinate systems) by replacing $\dot{\gamma}$ with $[\frac{1}{2}(\Delta:\Delta)]^{\frac{1}{2}}$, or, alternatively, τ with $[(\tau:\tau)]^{\frac{1}{2}}$, where $(\tau:\tau)$ is defined by the equation: $(\tau:\tau) = \Sigma_i\Sigma_j(\tau_{ij})^2$, in analogous fashion.

A point has now been reached when it is possible to use fundamental equations for the purpose of deriving expressions for the velocity and temperature profiles in simple flow geometries in terms of the parameters embodied in the fundamental equations, provided the boundary conditions are known and provided that certain reasonable simplifying assumptions are permissible.

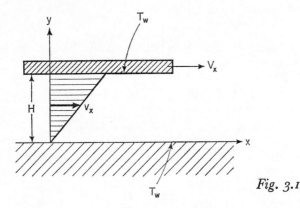

Fig. 3.1

Example 1

Here the simplest possible geometry (rectangular) is being considered as in flow between parallel plates. The liquid under shear obeys the power law, with power index $n = 1$ (the Newtonian case). The fluid volume is open to atmosphere and the hydrostatic pressure is therefore constant throughout the liquid. The plates are separated by a distance H which is small compared to the length and width of the plates. The lower plate is stationary whilst the upper plate moves at a constant speed V_x in the x-direction. Both plates are at a constant temperature T_w and it is assumed that all the properties of the liquid, including viscosity and density ρ are constant and that no body forces act upon the liquid (Fig. 3.1).

It is assumed that no flow occurs in the y-direction, or in the z-direction (perpendicular to the plane of the paper), so that flow is restricted to the x-direction only.

From these basic data it should be possible to calculate the temperature and the velocity of flow of the liquid at any point between the parallel plates in terms of H, V_x, T_w and η. Since the components of the velocity vector in the y and z directions are zero under our assumption, all the derivative terms of ∂v_y and ∂v_z are also zero. If the liquid is in steady laminar flow, any plane parallel to the plates will be characterised by the same velocity components v_x, so that the derivatives of v_x with respect to x and z must both be zero. The only nonzero component left (see Table V) is therefore $\partial v_x / \partial y$. The velocity vector $\mathbf{v}$ is therefore solely a function of the independent variable which defines the position of any given volume element within the gap between the plates, and this is the y-position co-ordinate.

The stress tensor $\boldsymbol{\tau}$ which, of course, arises from the velocity vector, will therefore likewise have only one component, namely τ_{yx} (all others being zero). The temperature generated as a result of the mechanical loss (friction) within the gap, is again clearly a function of the y-position co-ordinate for any one material with a known coefficient of thermal conductivity. This means that heat flux can occur in the y-direction only, so that the only nonzero component of the conductive heat flux vector $\mathbf{q}$ is q_y.

For this special case the left-hand side of the expansion of the momentum equation in rectangular co-ordinates of the x-component is zero and the only remaining term on the right-hand side is $\partial \tau_{yx} / \partial y$, so that the momentum equation (see Table III, p. 48) reduces to

Equation 3.1
$$\frac{\partial \tau_{yx}}{\partial y} = 0$$

As regards the energy equation (Table IV, p. 49), all the left-hand terms are zero on our assumptions, whilst of the right-hand side the only terms remaining are the second and the tenth out of a total of fifteen. The equation thus takes the form:

$$0 = -\left(\frac{\partial q_y}{\partial y}\right) + \tau_{xy}\left(\frac{\partial v_x}{\partial y}\right)$$

or

6

Equation 3.2
$$\tau_{xy}\left(\frac{\partial v_x}{\partial y}\right) = \left(\frac{\partial q_y}{\partial y}\right)$$

or in words:

The rate of loss of mechanical energy as generated heat (per unit volume) at any point is equal to the rate at which heat is lost by conduction at that point.

From Table I on p. 43, $q_y = -k \, \partial T/\partial y$.

Substituting in (3.2) we get:

$$\tau_{xy}\left(\frac{\partial v_x}{\partial y}\right) = -k \left(\frac{\partial^2 T}{\partial y^2}\right)$$

where k is the thermal conductivity of the liquid.

Now, τ is a symmetrical tensor, so that the subscripts of components are interchangeable and $\tau_{xy} = \tau_{yx}$, so that we can rewrite the last equation:

Equation 3.3
$$\tau_{yx}\left(\frac{\partial V_x}{\partial y}\right) = -k \left(\frac{\partial^2 T}{\partial y^2}\right)$$

The boundary conditions are:

Equation 3.4
$$v_x(y = 0) = 0 \qquad T(y = 0) = T_w$$
$$v_x(y = H) = V_x \qquad T(y = H) = T_w$$

the rheological equation which applies is the power-law equation, with n being equal to unity, so that

$$\tau = \eta \dot{\gamma}$$

The only nonzero component of the stress tensor is τ_{yx}. The strain rate, in terms of the velocity vector, consists of six shear and three normal force components. The normal force components $(\partial v_i/\partial i)$ are assumed to be zero (no elasticity). Since this is a case of simple shear, only one of the six shear components $\partial v_i/\partial j$ is nonzero. The only nonzero shear component is, of course, the one which arises from the movement relative to one another of the shear laminae in the planes parallel to the plates, as defined by the y-position co-ordinate $\partial v_x/\partial y$, so that the above equation can be written:

Equation 3.5
$$\tau_{yx} = \eta \left(\frac{\partial v_x}{\partial y}\right)$$

We now have three independent equations (3.1), (3.3) and (3.5) obtained from the momentum, energy and rheological equation respectively. To solve them it is necessary to carry out several integration operations which bring forth integration constants. In order to evaluate these integration constants use is made of the known boundary conditions as stated in Equations 3.4.

The series of operations involved is as follows:

(i) Integrate $\partial \tau_{yx}/\partial y = 0$ (Equation 3.1), which gives τ_{yx} directly:

Equation 3.6
$$\tau_{yx} = C_1$$

(ii) Substitute for τ_{yx} in Equation 3.5:

$$C_1 = \eta \frac{\partial v_x}{\partial y}$$

and integrate to give v_x:

Equation 3.7
$$v_x = \frac{C_1 y}{\eta} + C_2$$

(iii) Substitute for τ_{yx} and v_x in Equation 3.3:

$$C_1 \cdot \frac{C_1}{\eta} = -k \left(\frac{\partial^2 T}{\partial y^2} \right)$$

or

$$\frac{\partial^2 T}{\partial y^2} = -\frac{C_1^2}{\eta k}$$

which, on integrating twice, gives:

Equation 3.8
$$T = -\frac{C_1^2}{2\eta k} y^2 + C_3 y + C_4$$

where C_3 and C_4 are the respective integration constants.

(iv) Evaluate the integration constants from the known boundary conditions as expressed in Equations 3.4:

Since $C_1 = \eta \, (\partial v_x/\partial y)$, we can set $\partial y = H$, when ∂v_x becomes V_x and

Equation 3.3a
$$C_1 = \eta \frac{V_x}{H}$$

Substituting in (3.7) we get (setting $V_x = v_x$ and $y = H$):

Equation 3.3b
$$V_x = V_x + C_2$$

whence $C_2 = 0$.

Since $T = T_w$ when $y = 0$ it follows from Equation 3.8 that

Equation 3.3c
$$C_4 = T_w$$

Since $T = T_w$ also when $y = H$, substitution in (3.8) gives:

$$T_w = -\frac{C_1^2}{2\eta k} H^2 + C_3 H + T_w$$

or

$$C_3 = \frac{C_1^2}{2\eta k} \cdot H = \eta^2 \frac{\dfrac{V_x^2}{H^2}}{2\eta k} \cdot H$$

or

Equation 3.3d
$$C_3 = \frac{\eta V_x^2}{2kH}$$

(v) Introduce the expressions for the integration constants (Equations 3.3a, 3.3b, 3.3c, 3.3d) into Equations 3.6, 3.7 and 3.8:

Equation 3.9
$$\tau_{yx} = \eta \frac{V_x}{H}$$

Equation 3.10
$$v_x = y \frac{V_x}{H}$$

Equation 3.11
$$(T - T_w) = \frac{\eta}{2k} \left(\frac{V_x}{H}\right)^2 (Hy - y^2)$$

These final equations satisfy the equations of momentum energy and the rheological equation and also the specific boundary conditions of the flow geometry. The procedure is the same for any flow behaviour and any geometry, provided the appropriate form of the equations and boundary conditions are employed (see later).

Some interesting facts emerge from the inspection of Equations 3.9, 3.10 and 3.11. Equation 3.9 is perhaps not of *direct* importance in processing and can be left at this stage. According to Equation 3.10 the 'dimensionless velocity' v_x/V_x, when plotted against the position variable y/H, should give a 45° linear function through the origin (slope of unity) and this is in fact observable under suitably arranged experimental conditions.

Equation 3.11 can be rewritten:

$$(T - T_w) \frac{2k}{\eta V_x^2} = \frac{y}{H} \left(1 - \frac{y}{H}\right)$$

which shows that the left-hand side is a function solely of the positional variable y/H and that the function is parabolic with a maximum when $y/H = \frac{1}{2}$. It is, of course, obvious that the greatest temperature difference $(T - T_w)$ should be in the centre of the gap between the plates where the frictional shear heat cannot be dissipated as readily by conduction to the plates as elsewhere, but it does not follow *a priori* that the function should be parabolic.

Example 2

Again, we are considering an incompressible liquid in laminar flow which obeys the power law and which has a power index $n = 1$ (Newtonian), but this time we are using a long tube of circular cross section of radius R instead of two parallel plates. The tube wall is at a constant temperature T_w. The objective of the flow characterisation (as before) is to calculate the velocity and temperature profiles across the tube far enough from the ends of the tube so as to be unaffected by possible entrance and exit disturbances.

The co-ordinates are, of course, cylindrical with axes r, θ and z. Consideration of the main flow characteristics (with simplifying assumptions analogous to those made before) lead to the following realisations:

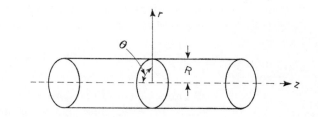

Fig. 3.2

(i) All derivatives of temperature T, velocity components v_i and shear stress components τ_{ij} with respect to time t and the axes θ and z are zero;

(ii) The velocity components v_θ and v_r are themselves zero;

(iii) div **v**, because of incompressibility, is also zero. Using the tables given earlier on, the equations of momentum and energy and the rheological equation therefore reduce to:

Equation 3.12
$$\frac{\partial P}{\partial z} = \frac{1}{r}\frac{\partial}{\partial r}(r\tau_{rz}) \quad \text{(Momentum)}$$

Equation 3.13
$$\frac{1}{r}\frac{\partial}{\partial r}(rq_r) = \tau_{rz}\left(\frac{\partial v_z}{\partial r}\right) \quad \text{(Energy)}$$

Equation 3.14
$$\tau_{rz} = \eta\left(\frac{\partial v_z}{\partial r}\right) \quad \text{(Rheological)}$$

Introducing the only component of the heat flux vector which is nonzero (q_r), given by:

$$q_r = -k\frac{\partial T}{\partial r}$$

(see Table I, p. 43), and Equation 3.14 into 3.12 and 3.13 respectively, we obtain:

Equation 3.15

$$\frac{\partial P}{\partial z} = \frac{\eta}{r} \frac{\partial}{\partial r} \left[r \left(\frac{\partial v_z}{\partial r} \right) \right]$$

and

Equation 3.16

$$-\frac{k}{r} \frac{\partial}{\partial r} \left[r \left(\frac{\partial T}{\partial r} \right) \right] = \eta \left(\frac{\partial v_z}{\partial r} \right)^2$$

The boundary conditions are given by:

$$\begin{cases} (a) \ldots v_z(r = R) = 0 & (c) \ldots \frac{\partial v_z}{\partial r} (r = 0) = 0 \\ \\ (b) \ldots T(r = R) = T_w & (d) \ldots \frac{\partial T}{\partial r} (r = 0) = 0 \end{cases}$$

Note that the left-hand side of the momentum equation (3.15) does not depend on r and that the equation can therefore be integrated directly, giving, on rearrangement:

Equation 3.17

$$\frac{\partial v_z}{\partial r} = \frac{r}{2\eta} \frac{\partial P}{\partial z} + \frac{C_1}{r}$$

Because of boundary condition (c), the integration constant C_1 must be zero. The velocity profile is obtained from (3.17) upon second integration which gives:

Equation 3.18

$$v_z = \frac{r^2}{4\eta} \frac{\partial P}{\partial z} + C_2$$

The integration constant C_2 follows from boundary condition (a), which when introduced into (3.18), shows that

$$C_2 = -\frac{R^2}{4\eta} \frac{\partial P}{\partial z}$$

so that (3.18) becomes:

Equation 3.19

$$v_z = -\frac{R^2}{4\eta} \left(\frac{\partial P}{\partial z} \right) \left[1 - \frac{r^2}{R^2} \right]$$

(The negative sign indicates that the liquid flows in the direction of decreasing pressure.)

The temperature profile is obtained by introducing $\partial v_z / \partial r$ as given by Equation 3.17 into Equation 3.16:

$$-\frac{k}{r} \frac{\partial}{\partial r} \left[r \left(\frac{\partial T}{\partial r} \right) \right] = \eta \left[\frac{r}{2\eta} \left(\frac{\partial P}{\partial z} \right) \right]^2$$

Equation 3.16a

$$= \frac{r^2}{4\eta} \left(\frac{\partial P}{\partial z} \right)^2$$

The first integration of Equation 3.16a so obtained gives:

$$-\frac{k}{r} \cdot r\left(\frac{\partial T}{\partial r}\right) = \frac{r^3}{16\eta}\left(\frac{\partial P}{\partial z}\right)^2 + \frac{C_3}{r}$$

or

Equation 3.16b

$$\left(\frac{\partial T}{\partial r}\right) = -\frac{r^3}{16\eta k} + \frac{C^3}{r}$$

Boundary condition (d), when introduced into (3.16b), shows that the integration constant C_3 is zero.

The second integration of (3.16b) gives:

Equation 3.20

$$T = -\frac{r^4}{64\eta k}\left(\frac{\partial P}{\partial z}\right)^2 + C_4$$

Boundary condition (b) is introduced into (3.20), showing that the integration constant C_4 is equal to:

$$\frac{R^4}{64\eta k}\left(\frac{\partial P}{\partial z}\right)^2$$

so that (3.20) becomes:

$$T - T_w = \frac{1}{64\eta k}\left(\frac{\partial P}{\partial z}\right)^2 (R^4 - r^4)$$

or

Equation 3.21

$$T - T_w = \frac{R^4}{64\eta k}\left(\frac{\partial P}{\partial z}\right)\left[1 - \left(\frac{r}{R}\right)^4\right]$$

Again, some interesting facts emerge from a contemplation of the velocity equation (3.19) and the temperature difference equation (3.21):

Firstly, the velocity at the tube centre v_0 is obtained by setting $r = 0$ in Equation 3.19, which then yields:

Equation 3.22

$$v_0 = -\frac{R^2}{4\eta}\left(\frac{\partial P}{\partial z}\right), \text{ so that}$$

Equation 3.23

$$v_0^2 = \frac{R^4}{16\eta^2}\left(\frac{\partial P}{\partial z}\right)^2$$

Secondly, by setting $r = 0$ in Equation 3.21, T in that equation becomes T_0 (the temperature at the centre of the tube) giving:

Equation 3.24

$$T_0 - T_w = \frac{R^4}{64\eta k}\left(\frac{\partial P}{\partial z}\right)^2$$

On comparing the right-hand sides of Equations 3.23 and 3.24, it is seen that the latter differs from the former by a factor of $\eta/4k$, that is to say, that

Equation 3.25
$$T_0 - T_w = \frac{\eta v_0^2}{4k}$$

Thirdly, it follows directly from (3.22), that

Equation 3.26
$$\left(\frac{\partial P}{\partial z}\right) = -\frac{4\eta v_0}{R^2}$$

Lastly, Equations 3.19 and 3.20 can be made dimensionless by dividing them by v_0 and $(T_0 - T_w)$ respectively, as given by (3.22) and (3.24), resulting in:

Equation 3.27
$$\frac{v_z}{v_0} = 1 - \left(\frac{r}{R}\right)^2$$

and

Equation 3.28
$$\frac{T - T_w}{T_0 - T_w} = 1 - \left(\frac{r}{R}\right)^4$$

These equations are of considerable usefulness and surprising simplicity.

It is obvious that the velocity and temperature profiles are both parabolic between the limits of $+R$ and $-R$ and that the maximum for both v_z and $(T - T_w)$ is at the tube centre. But whilst the velocity profile is a square parabola, the temperature profile is a parabola of the fourth power, so that the rate of temperature increase towards the centre from the tube wall is much greater than the rate of velocity increase.

Units

When evaluating actual flow problems it is necessary to use units consistently. Supposing we use cm, g-mass, sec, poise and °C—these may subsequently have to be converted to, say, psi (lb-force per sq inch) per inch die length (which is often convenient). In order to effect the conversion, the result of the calculation in absolute units is multiplied by the following factors:

(i) 10^{-3} (g-mass to kg-mass).

(ii) $\dfrac{1}{981}$ (kg-mass to kg-force).

(iii) 14·2 (kg-force per cm² to lb-force per sq inch).

(iv) 2·54 (cm⁻¹ length to inch⁻¹ length).

This may be lumped together into a constant C, where

$$C = 3 \cdot 68 \times 10^{-5}$$

Note also that k (which is usually given in cals cm^{-1} sec^{-1} (°C)$^{-1}$) must be converted into the mechanical equivalent of heat in absolute units and must therefore be multiplied by

$$4 \cdot 2 \times 10^7$$

We are now in a position to show how a *numerical example* may be solved:

Example 2a

A polymer melt with power index $n = 1$ has a viscosity of 10^4 poise at 170°C and flows through a cylindrical die of circular cross section (diameter = 4 mm) at a rate of 10 cm per second, the flow being laminar. Assuming that there are neither time dependent nor entrance effects, and assuming further that the simplifications which have led to Equations 3.19 and 3.21 above are permissible:

(i) What is the pressure drop per inch of axial direction at the die centre in psi?

(ii) What is the temperature at the die centre if the wall temperature is 170°C and if the thermal conductivity is 10^{-3} cal cm^{-1} sec^{-1} (°C)$^{-1}$?

(iii) How far from the centre of the die, to the nearest 0·01 mm, will the melt temperature be just 172°C?

Answer

(i) Set $r = 0$, then

$$v_z = v_0 = -\frac{R^2}{4\eta}\left(\frac{\partial P}{\partial z}\right)$$

$$\frac{\partial P}{\partial z} = -\frac{4v_0\eta}{R^2} = -\frac{4 \times 10 \times 10^4}{(0 \cdot 2)^2} = -10^7 \text{ dynes cm}^{-2}\text{ cm}^{-1}$$

$$= -10^7 \times 3 \cdot 68 \times 10^{-5} = \underline{\underline{-368 \text{ psi}}} \text{ (lb-force in}^{-2}\text{ in}^{-1}\text{)}$$

The pressure drop in the die is 368 psi per inch length.

(ii) According to Equation 3.23

$$T_0 - T_w = \frac{\eta v_0^2}{4k}$$

Since v_0 is the greatest velocity across the profile of the die, this is obviously the quoted extrusion rate (no draw being applied) of the material. Therefore,

$$T_0 = 170 + \frac{10^2 \times 10^4}{4 \times 10^{-3} \times 1\cdot 4 \times 10^7} = 170 + \frac{1,000}{16\cdot 8} = \underline{\underline{176°C}}$$

The temperature at the die centre will be 176°C, 6°C greater than at the wall.

(iii) According to Equation 3.28

$$\frac{T - T_w}{T_0 - T_w} = 1 - \left(\frac{r}{R}\right)^4$$

Let $T - T_w = 2$, then

$$\frac{2}{6} = 1 - \left(\frac{r}{R}\right)^4 \quad \text{or} \quad r = R \sqrt[4]{\frac{2}{3}} = R \times 0\cdot 9 = \underline{\underline{0\cdot 18 \text{ cm}}}$$

Example 3

In this, the final example of this genre, we are using exactly the same assumptions and geometry as in Example 2, with the sole exception that the liquid this time obeys the power law in which the power index $n \neq 1$.

The momentum and energy equations are of the same form as before and give the following intermediates before integration:

Equation 3.29
$$\frac{\partial P}{\partial z} = \frac{1}{r}\frac{\partial}{\partial r}\left[r\eta \left(\frac{\partial v_z}{\partial r}\right)\right]$$

and

Equation 3.30
$$-\frac{k}{r}\frac{\partial}{\partial r}\left[r\left(\frac{\partial T}{\partial r}\right)\right] = \eta \left(\frac{\partial v_z}{\partial r}\right)^2$$

corresponding to Equations 3.15 and 3.16 of Example 2, but differing from them in that η is no longer the constant viscosity of a Newtonian, but the shear rate dependent viscosity of a power-law liquid where $n \neq 1$.

The shear rate dependent viscosity is given by the power law which is conveniently written in the form

$$\eta = \eta^0 \left(\frac{\dot\gamma}{\dot\gamma^0}\right)^{n-1}$$

(see p. 51), which for $\dot\gamma^0 = 1$ sec^{-1} reduces to:

Equation 3.31
$$\eta = \eta^0 \dot\gamma^{n-1}$$

where η^0 is the standard viscosity at the standard reference shear rate of $\dot\gamma^0 = 1$ sec^{-1}.

Introducing Equation 3.31 into Equation 3.29, we get:

Equation 3.32
$$\frac{\partial P}{\partial z} = \frac{\eta^0}{r}\frac{\partial}{\partial r}(r\dot\gamma^n)$$

which, on integration with respect to r becomes:

$$\dot\gamma^n = \left(\frac{\partial v_z}{\partial r}\right)^n = \frac{r}{2\eta^0}\left(\frac{\partial P}{\partial z}\right) - \frac{C_1}{r}$$

Because the shear stress τ_{rz} at the centre of the tube is zero, therefore C_1 must be zero. Taking the nth root and putting $C_1 = 0$ we obtain:

Equation 3.33
$$\dot\gamma = \frac{\partial v_z}{\partial r} = \left[\frac{r}{2\eta^0}\left(\frac{\partial P}{\partial z}\right)\right]^{1/n}$$

which on carrying out a second integration with respect to r gives:

Equation 3.34
$$v_z = \frac{n}{n+1}\, r^{(n+1)/n}\left[\frac{1}{2\eta^0}\left(\frac{\partial P}{\partial z}\right)\right]^{1/n} + C_2$$

The integration constant C_2 is obtained from the boundary condition:

Equation 3.35
$$v_z(r = R) = 0$$

so that one finally obtains the equation:

Equation 3.36
$$v_z = -\frac{n}{n+1} R^{(n+1)/n}\left[\frac{1}{2\eta^0}\left(\frac{\partial P}{\partial z}\right)\right]^{1/n}\left[1 - \left(\frac{r}{R}\right)^{(n+1)/n}\right]$$

At the centre of the tube $v_z = v_0$ and $r = 0$, so that Equation 3.36 becomes:

Equation 3.37
$$v_0 = -\frac{n}{n+1} R^{(n+1)/n}\left[\frac{1}{2\eta^0}\left(\frac{\partial P}{\partial z}\right)\right]^{1/n}$$

Equation 3.36 can be written in dimensionless form:

Equation 3.38
$$\frac{v_z}{v_0} = \left[1 - \left(\frac{r}{R}\right)^{(n+1)/n}\right]$$

Note that for $n = 1$, Equation 3.38 reduces to the Newtonian case, identical with Equation 3.27 of Example 2.

To obtain the temperature profile we introduce Equations 3.31 and 3.33 into the energy Equation 3.30 and integrate with respect to r to obtain:

Equation 3.39
$$\frac{\partial T}{\partial r} = -\frac{\eta^0}{k}\cdot\frac{n}{3n+1}\left[\frac{1}{2\eta^0}\left(\frac{\partial P}{\partial z}\right)\right]^{(n+1)/n} r^{(2n+1)/n} + \frac{C_3}{r}$$

Using the boundary condition that at the tube centre ($r = 0$) $\partial T/\partial r$ is zero, substitution in Equation 3.39 shows that C_3 must be zero. Integrating a second time with respect to r we obtain:

Equation 3.40 $$T = -\frac{\eta^0}{k}\left(\frac{n}{3n+1}\right)^2\left[\frac{1}{2\eta^0}\left(\frac{\partial P}{\partial z}\right)\right]^{(n+1)/n} r^{(3n+1)/n} + C_4$$

The integration constant C_4 follows from the boundary condition that the temperature at radius $r = R$ is equal to T_w and Equation 3.40 therefore becomes:

Equation 3.41 $$T - T_w = \frac{\eta^0}{k}\left(\frac{n}{3n+1}\right)^2 R^{(3n+1)/n}\left[\frac{1}{2\eta^0}\left(\frac{\partial P}{\partial z}\right)\right]^{(n+1)/n}$$
$$\left[1 - \left(\frac{r}{R}\right)^{3(n+1)/n}\right]$$

The difference in the liquid temperature between the centre of the tube and the wall is obtained from the boundary condition that at $r = 0$ the temperature is T_0, so that, by putting $r = 0$, Equation 3.41 becomes:

Equation 3.42 $$T_0 - T_w = \frac{\eta^0}{k}\left(\frac{n}{3n+1}\right)^2 R^{(3n+1)/n}\left[\frac{1}{2\eta^0}\left(\frac{\partial P}{\partial z}\right)\right]^{(n+1)/n}$$

and by dividing Equation 3.41 by Equation 3.42 the dimensionless form

Equation 3.43 $$\frac{T - T_w}{T_0 - T_w} = 1 - \left(\frac{r}{R}\right)^{(3n+1)/n}$$

is finally obtained—again a simple and elegant result.

Note that Equation 3.43 becomes identical with (3.28) of Example 2, if $n = 1$, when it reduces to the Newtonian case.

———

The temperature difference across the die profile is obviously critically dependent on the magnitude of the power index n. It can be substantial under conditions where the shear stress is high. This is certainly the case under extrusion conditions, but even more so in the case of polymer melt passing through the runners and pinpoint gates of an injection moulding system. Indeed it is not at all uncommon that the temperature increase should be so large that thermal decomposition of the material occurs and mouldings show scorch marks. The remedy is obvious:

Either the viscosity is reduced by one of the following means:

(i) Selection of a lower viscosity (i.e. a lower molecular weight) grade of the same material. (This will be easier to mould but the mechanical strength of the moulding will be reduced.)

(ii) Increasing the melt temperature (paradoxically!).

(iii) Widening of gate diameter and/or runner diameters, *or* reducing the pressure drop. This always has a beneficial effect on extrudate quality because it reduces residual stresses, increases the dwell time in the die and thus minimises the risk of flow defects, but it naturally reduces the output rate. In moulding it is generally desirable to work at the lowest possible injection pressures in order to obtain the best quality moulding but in order to be able to fill the mould completely, the pressure must nevertheless be of a high order—about ten times as high as in extrusion. Naturally, it helps to fill the mould if the mould temperature is kept as high as possible, but whilst this assists in mould filling and minimises moulding stresses, it also increases cycle times and so reduces output.

It may be possible to increase the thermal conductivity of the melt by incorporating finely divided fillers with a specific conductivity greater than that of the polymer melt (e.g. metal powders), but this has not so far been regarded as a practical means of reducing temperature differences, because the changes in the properties of the extrudate or moulding (as the case may be) will obviously be rather extreme as a result of such inclusions.

In the calculations which have been carried out in this chapter it was assumed that the viscosity is constant across the die at any given shear stress. But this is patently incorrect since the temperature changes will cause corresponding viscosity changes. These may be ignored as a first approximation, but if the temperature difference across the die is more than (literally) one or two degrees centigrade the viscosity changes of polymer melts can be quite substantial and it is advisable to carry out a series of approximations.

In order, however, to deal with this aspect properly, it is necessary to consider in greater detail the temperature dependence of viscous flow and its activation energy. This will be the subject of Chapter 4.

4 The temperature, pressure, shear rate and molecular weight dependence of viscous flow

Simple mechanisms in elasticity and flow, with reference to crystallinity. Energy considerations—the Arrhenius equation and its modification. Diffusion. The free-volume concept as the basis for explaining temperature and pressure dependence of flow. The effect of shear rate; shear-induced crystallinity. The flow unit in chainlike molecular structures and its relation to molecular weight. Temperature/viscosity relationships in non-Newtonian liquids—activation energies at constant shear rate and at constant shear stress and their relation to the power index of power-law liquids; their use in predicting melt viscosities of extrusion materials. Determination of activation energies.

Simple mechanisms in elasticity and flow

BEFORE concerning ourselves with the mechanism of flow in liquids, it is profitable to consider first the mechanism of deformation in a more highly ordered system. The most highly ordered system for any assemblage of constituent units is found in the crystalline solid state. There are, of course, varying degrees of disorder even in the solid crystalline state, because some crystals are less than perfect. Even when they *are* perfect, however, crystals can be organised with reference to varying numbers of axes and planes of symmetry. The crystal structure possessing the largest number of axes and planes of symmetry is the orthorhombic system, so that one would expect orthorhombic crystals to be in a state of ideal order. This, however, is still far from the truth.

The forces of cohesion between the units making up the structure—however strong they may be—are in dynamic equilibrium with opposing forces. This means that whilst there may be a high probability of finding unit B (at locus Y) relative to unit A (at locus X) there is no absolute certainty of finding it in this locus *at all times*. Indeed, if the position of unit B is plotted against time in such a way that the extremes of the possible distances from locus Y are denoted by $(Y + dY)$ and $(Y - dY)$ respectively (Fig. 4.1) a sinusoidal curve will be obtained which shows that the probability of finding B *exactly* at locus Y is infinitely small, although we can be certain to find B within a sphere of radius dY at all times,

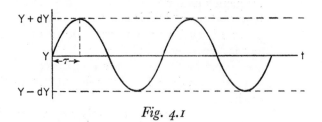

Fig. 4.1

dY being the amplitude of the sinusoidal function. The amplitude dY depends on an intrinsic time scale τ of unit B which is in turn a function of the intrinsic energy of B and its mass. B is energised by increasing the temperature from the absolute zero. Only at the absolute zero will the kinetic energy of B be zero and only in that case will dY be zero. In that special case, of course, there will be absolute certainty that B will be found at locus Y exactly.

The amount of kinetic energy necessary to make B commence its sinusoidal evolutions will depend on its mass. This makes it theoretically possible to determine the mass of unit B by measuring the amount of heat which must be supplied to B to force it to leave its ground state position and by then converting that heat to its mechanical equivalent. Unless the temperature of our 'perfect' and maximally symmetric crystal is below the critical temperature at which B begins to oscillate (three-dimensionally) around Y, the crystal will be less than perfect. If one imagined a time exposure photograph to be taken above that critical temperature all the units making up the crystal structure will appear to be out of focus. The greater the mass of the units, the higher will be the temperature at which the units can still remain at rest (or the greater will be the number of quanta of energy which must be supplied to force them out of the ground state). This, then gives us a reference point in terms of temperature for the commencement of deformational processes. Once the process of dynamic oscillation starts, however, its frequency will increase linearly and its amplitude will increase exponentially with increasing temperature. These principles are embodied in the Arrhenius equation which applies to all rate processes including viscous and elastic deformation:

Equation 4.1 $$\text{Rate of process} = \frac{\text{nett change}}{\text{unit time}} = \frac{dD}{dt} = A\,e^{-E/RT}$$

where R is the gas constant, E is the activation energy of the process and A is a frequency term.

The nett change dD therefore comprises the product of a frequency factor $A dt$ and an energy factor $e^{-E/RT}$. At constant temperature $e^{-1/RT}$ is constant. E is also constant because the process itself involves work done in its accomplishment (such as the breaking of existing or the creation of new geometrical conformations between the constituent units) and the amount of such work done is strictly dependent on the observed degree of change.

The Arrhenius equation is experimentally verifiable over limited ranges of temperature by using it in its logarithmic form, when

Equation 4.2
$$\ln \frac{dD}{dt} = A' - \frac{E}{RT}$$

so that a plot of $\ln dD/dt$ vs $1/T$ should give a straight line with slope A' and intercept $-E/R$. Over wide temperature ranges, however, the function shows substantial curvature, implying that E is not a constant which is contrary to the definition of E. This difficulty is solved in the following way:

The exponential $-E/RT$ can be written $-E/R(T - T_0)$, where T_0 is the absolute zero and $(T - T_0)$ reflects the degree to which the system has been energised. From what has been said before, however, it is obvious that not *all* that energy is being converted into kinetic energy or its chemical or mechanical equivalent. It is not therefore available *in toto* for the purpose of effecting a change in state D. Part of the energy is expended in overcoming the inertia due to the mass of the units in their ground state. That part is indicated by the true reference temperature T_r, the critical temperature below which a change in state D cannot be effected. The original Arrhenius plot makes no allowance for the fact that the energy term E is the sum of the energy E_r necessary to force the system out of its ground state and the true activation energy E_a which alone is relevant to effecting a change in state D. If

Equation 4.3
$$E = E_r + E_a$$

then it follows that

Equation 4.4
$$E_a = E - E_r$$

Since E corresponds to a temperature change $(T - T_0)$, E_r corresponds to a temperature change $(T_r - T_0)$ and E_a must correspond to a temperature change $(T - T_r)$.

Replacing the energies by the corresponding temperature changes, in Equation 4.4, we get

$$(T - T_r) = (T - T_0) - (T_r - T_0)$$

which is an obvious identity.

It follows that T_0 is irrelevant and that the reference temperature which yields a truly constant and meaningful activation energy E_a is T_r.

To test the soundness of this theory it is merely necessary to rewrite the Arrhenius equation:

Equation 4.5
$$\frac{\mathrm{d}D}{\mathrm{d}t} = A\,e^{-E_a/R(T - T_r)}$$

and to see whether the log form of this equation gives a straight line from the observed data for a given process at some temperature T_r which can be determined by trial and error. The fact that the Arrhenius equation has served chemists so well in its unmodified form is due to the fortuitous circumstance that the processes studied by classical chemistry required very low E_r values and that T_r is therefore fairly close to T_0. In other words, the masses of the constituent parts of structures (crystalline or otherwise) were small—usually of atomic dimensions.

But in processes involving the deformation and flow of *polymers* T_r can be very high indeed and may actually only be a few tens of degrees below the static glass transition point T_g. This is due to the fact that the masses of the constituent units of deformation and flow in polymers are very much larger than those corresponding to mere atoms, groups of atoms or even repeating units. We shall return to this presently.

It should be stated, however, that there may still be disturbing influences present even in an apparently perfect crystal at temperatures below T_r, because of imperfections within the constituent units themselves. Firstly, assuming even that the constituent units are atoms of the same chemical nature (which is not necessarily the case), then these atoms may still be isotopes and will therefore vary in nuclear mass, so that we can have various isotopes randomly distributed in the crystal, each individual one with its own characteristic T_r. This will result in a whole spectrum of species of lattice cells.

Secondly, assuming that even all the isotopes are identical, it is still possible for different types of nuclear magnetic resonance, electron magnetic resonance, spins and spin couplings to occur.

71

7

All this goes to show that an ideal crystal with no disturbance must be a very rare bird indeed and it is not surprising that it has never been detected. It can, of course, be proved thermodynamically that an ideal crystal, that is to say a crystal with no disturbance or with zero entropy, can only exist at absolute zero temperature, a temperature which is not experimentally realisable.

The more complex a unit of the structure is, the greater will therefore be E_r, not only because of the larger mass directly, but also because the unit itself can absorb energy in a variety of modes within itself (rotational, translational or vibrational) whilst still remaining in an apparent ground state as far as an observer located on a neighbouring unit is concerned. To investigate this microcosm of internal changes *within* a unit it is necessary to eliminate all gross external effects such as those caused by mechanical manipulation of the structure in bulk and to have recourse to sophisticated methods designed to excite specifically only one kind of response. This is accomplished by excitation with ultraviolet, visible, infra-red, X-ray, electrical and magnetic devices and the subsequent interpretation of the resulting spectra.

All this has been introduced here in order to serve as an aid to conceptualisation, but it also serves to illustrate another interesting aspect of material science.

It is clear that mechanical forces capable of affecting the relative positions of units may also be of sufficient magnitude to affect the existing dynamic equilibrium within the unit itself. Mechanical deformation (whether stored or dissipated) may therefore result in producing secondary effects, such as luminescence, electrical polarisation, chemical breakdown or changes in magnetic behaviour, which can otherwise be caused only by subjecting the material specifically to the respective selective type of excitation. This means that a mechanical spectrum can never be expected to be 'pure', since it always has the fine-structural overtones which properly belong to another form of excitation. However, in general the overtones are of a much lower order of magnitude than the mechanical excitation itself and they may therefore be disregarded unless one is concerned with research on fine structure.

The mechanical forces necessary to cause certain effects may be classified as follows:

(*a*) Those necessary to deform valence bonds and bond angles between the constituent atoms of a molecule. They give rise to a modulus of the order of 10^{12} dyne cm^{-2}.

(*b*) Those necessary to displace molecules relative to one another against secondary valence or Van der Waals' forces, giving rise to a modulus of the order of 10^9 to 10^{11} dyne cm^{-2}.

(c) Those necessary to overcome the rather weak London forces, such as the forces of attraction existing between saturated aliphatic hydrocarbons, giving rise to a modulus of the order of 10^6 to 10^8 dyne cm^{-2}.

The 'softest' of these responses will dominate the gross mechanical behaviour since it constitutes the 'weak link' amongst the cohesive forces. If the weak link is of the London forces type (as in polythene), then the modulus will be low, if it is of the Van der Waals' force type (as the hydrogen bonding in nylons) then the modulus will be high; if we are dealing with a single crystal, then primary valence bonds and bond angles are involved and the modulus will be at its highest, but obviously objects made in plastics cannot consist of single crystals, so that moduli of the order of 10^{12} dyne cm^{-2} can never be realised in plastics.

More complex mechanisms of flow

To return to T_r, the critical reference temperature for relaxation processes: the idea of the existence of such a temperature is not new. Miller [1] has shown that published viscosity data for polystyrene and PIB conform to a modified Arrhenius equation:

Equation 4.6
$$\eta = A\, e^{B/T - T_r}$$

From this equation and from another treatment of the same data based on Doolittle's 'free volume' concept [2] which subsequently received attention from other workers [3, 4, 5, 6] the reference point T_r was derived. Both approaches show that the reference point lies on a specific volume/temperature curve which is extrapolated linearly into the glass region of the polymers concerned and is therefore associated with a corresponding volume V_r. This work stemmed from the observed fact that in relaxation processes in liquids (which are rate processes and for which the Arrhenius equation should therefore apply) the simple equation

Equation 4.7
$$\tau = A\, e^{-E/RT}$$

does not hold adequately, especially around the static glass

73

transition point T_g. A good fit is, however, obtained by modifying the equation to

Equation 4.8 $$\tau = A\, e^{-B/T - T_r}$$

where T_r is a temperature well below T_g (Cole).

When Doolittle studied the viscosity behaviour of n-alkanes he formulated the 'free volume' concept:

Equation 4.9 $$\eta = A\, e^{BV_r/V_f}$$

where V_r = the 'occupied volume' (i.e. the volume occupied by the molecules themselves), and

V_f = the 'free volume' (i.e. the difference between the volume occupied by the material mass and V_r).

Guttmann and Simmons showed that Doolittle's n-alkane viscosity data could equally be represented by the modified Arrhenius equation and Cohen and Turnbull demonstrated the equivalence of the free-volume equation and the modified Arrhenius equation. Cohen and Turnbull's essential points are that:

(i) The free volume V_f is zero at T_r.

(ii) V_f increases linearly with temperature above T_r.

The experimental determination of the point $(V_r,\ T_r)$ was done in two ways:

1. Williams calculated the free volume from experimentally determinable quantities such as η, T_g, the specific volume at T_g (V_g) and the thermal coefficient of expansion dV/dT of the melt, employing data which Fox and Flory had obtained for twelve polystyrene fractions. He substituted the value in Doolittle's Equation 4.9 and used a programmed computer to obtain the best values for A, B and V_r which were:

$$V_r = 0 \cdot 937 \pm 0 \cdot 003 \text{ ccs gm}^{-1}$$
$$\log A = -3 \cdot 47$$
$$B = 0 \cdot 91$$

2. Miller started by using that coefficient of thermal expansion which applies below T_g down to T_r, taking advantage of the fact that dV/dT is also linear in that region. Using the reference temperature T_r, the total volume is given by the sum of the occupied volume V_r and the free volume V_f, where

Equation 4.10 $$V_f = (T - T_r)\frac{dV}{dT}$$

where V_f represents the free volume at temperature T and T

and dV/dT have been defined above.

Dividing Equation 4.10 by V_r we get

Equation 4.11
$$\frac{V_f}{V_r} = \frac{dV}{dT}(T - T_r)\frac{1}{V_r}$$

where the coefficient of $(T - T_r)$ is obviously a constant which Miller denotes as α. Taking reciprocals of (4.11) we obtain

Equation 4.12
$$\frac{V_r}{V_f} = \frac{1}{\alpha(T - T_r)}$$

and introducing this into Doolittle's equation (4.9) the result is:

Equation 4.13
$$\eta = A\,e^{B/\alpha(T - T_r)}$$

the modified Arrhenius equation, if B/α is identified with $-E/R$ which it obviously can be, since B/α is a material constant.

All that is necessary now is to take viscosity data from the literature, e.g. Fox and Flory's data for a polystyrene fraction of molecular weight 1,675 ($T_g = 40°C$) and to plot $\ln \eta$ against $1/T - T_r$ (Fig. 4.2) using trial values for T_r until a straight line is obtained the slope of which will be B/α, or $(-E/R)$. With the exponential term defined, A follows directly from the Doolittle equation. Curves (a) to (e) were obtained (Fig. 4.2) by assuming the following values of temperature for T_r:

(a) −273°C (absolute zero).
(b) 0°C.
(c) 5°C.
(d) 10°C.
(e) 40°C (T_g for the material).

The linearity of (c) denotes that 5° is the correct value for T_r.

It is noteworthy that substantial deviations from linearity occur if T_r is taken only a few degrees either side of the true value, and that a hopeless fit is obtained if T_r is assumed to be either the absolute zero or T_g.

The resulting value for A was −3·45 in excellent agreement with Williams's value of −3·47.

If the procedure is repeated with other polystyrene fractions of known molecular weight and viscosity data, it is possible to derive viscosity–temperature–molecular weight relationships. This was done and it was seen that above a molecular weight of about 35,000 further increases in molecular weight had comparatively little influence on the viscosity of polystyrene melts at any one temperature. Both T_r and T_g increase with molecular weight to a limiting value. This is to be expected, at least as far as T_r is concerned because T_r will increase as the mass of the flow unit increases until the molecular weight

75

reaches a value when whole molecules can no longer be the flow units due to entanglements, so that chain segments of a certain average length become the flow units instead. These flow units will obviously not increase in length, however much longer the complete polymer chain may be.

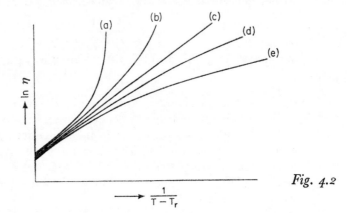

Fig. 4.2

The slope of the modified Arrhenius plot gives the true activation energy E_r. At higher temperatures, that is to say, at lower values of $1/T$ (Fig. 4.2) the apparent activation energy E_a obtained from the unmodified Arrhenius plot approximated to E_r which constitutes the asymptotic value. With decreasing temperature, however, E_a increases exponentially and attains extremely high values at T_g.

Miller has carried out a similar analysis of flow data for PIB with equally satisfactory correlations.

Miller's work and that of other workers have thus supplied first class experimental and corroborative evidence for the ideas derived from the extremely simple first principles as set out in the earlier part of this chapter which were based on crystal lattice considerations.

Diffusion

The question which remains to be answered is whether it is permissible to apply the considerations based on a crystal lattice to viscous flow in polymer melts. There can be no doubt that the answer to that question is in the affirmative:

The molecules in a liquid are in essentially crystal-like surroundings, but the 'crystals' are highly imperfect. A given unit (shaded in Fig. 4.3 below) may have a choice of two energetically equal positions of which it can only fill one at a time, leaving a 'hole' in the other (blank in Fig. 4.3). Now, the two positions are separated by the restricted opening between the arrowed units, so that the unit, in jumping from its presently occupied place into the hole, will need to do work

Fig. 4.3

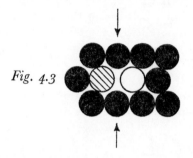

against the arrowed units. It will have to surmount an energy barrier. If it possesses sufficient energy it will do so with greater or lesser ease. The unit can be energised by supplying heat to the system and the heat will also cause the system to expand. This will widen the bottleneck between the two spaces available for the shaded unit. Once the energy hump is sufficiently small the energised unit will jump fairly freely between its available positions and other units (in reality indistinguishable from the first) will likewise jump to and fro into the positions vacated by their neighbours. This is essentially a diffusion process controlled by the frictional forces (i.e. the viscosity) between units. Since the gaps open up as the temperature increases, the friction becomes less and the viscosity decreases. This is true for both the solid and the liquid state, the differences between the two states being essentially due to the large differences in the order of magnitude of internal friction respectively.

The rate of molecular diffusion is equal to the number of jumps per second. It is inversely proportional to the frictional constant η and is related exponentially to the true activation energy of the process:

Number of jumps per unit time

$$= \text{Number of available holes} \times \frac{1}{\eta} \times e^{-E_r/k(T - T_r)}$$

or

Equation 4.14 $$\eta = \frac{\text{Number of available holes}}{\text{Number of jumps per unit time}} e^{-E_r/k(T - T_r)}$$

where k is the Boltzmann constant.*

* $k = R/N$, where $N = $ the number of molecules per mole.

It is clear that the fraction on the right-hand side of Equation 4.14 is equal to the constant A in the Arrhenius equation and that A is therefore a frequency constant. The jumps in a body at rest are equally frequent in all directions, so there is no nett deformation or flow. In a liquid the number of available holes is large because of the magnitude of the lattice imperfections compared to that prevailing in the solid state. But the number of jumps per unit time will be very much larger still and A will therefore be small.

If an external stress is applied which pushes the units in a preferred direction then the potential energy valleys will be depressed in the flow direction (Fig. 4.4), so that the activation

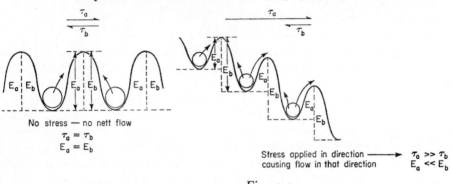

Fig. 4.4

energy E_b which must be imparted to make the units move in the reverse direction becomes so great that the probability of units surmounting this energy hump and moving against the applied stress becomes negligibly small, although in rigorous treatment allowance should be made for it.

Pressure dependence

For a short time interval a stressed liquid will behave like an elastically deformed solid, but as the rate process of jumps gets under way, the molecules gain kinetic energy ($E_b - E_a$ per jump) and since this cannot be stored, it must be dissipated as heat. The higher the viscosity, the more important becomes the elastic response. Batschinski has proposed the equation

Equation 4.15
$$\eta = \frac{I}{C(V_l - V_s)}$$

where C is a constant for any one material and V_l and V_s are the molar volumes of the units in the liquid and solid state respectively. $(V_l - V_s)$ therefore represents the increase in free volume during the solid/liquid transition.

Batschinski's equation is, however, open to criticism in the light of what has been developed a little earlier on in this chapter.

The equation implies that at the solid/liquid transition when $V_l = V_s$ the viscosity is infinity. This is obviously not so, because viscous flow in solids shows itself as creep; large though it may be, it is most certainly finite and measurable. Clearly, the relationship will be improved by writing V_r for V_s, and V for V_l so that (4.15) becomes

Equation 4.16
$$\eta = \frac{1}{C(V - V_r)}$$

where V_r is the volume occupied by the units proper (i.e. the volume occupied by the material mass at the reference temperature T_r) and where V is the volume of the material mass at any temperature T above T_r, irrespective of whether this falls into the solid or liquid state region. $(V - V_r)$ is, of course, Doolittle's free volume V_f and the equation therefore reduces further to:

Equation 4.17
$$\eta = \frac{C}{V_f}$$

which is a mathematical statement that viscosity is inversely proportional to the free volume.

This equation however, when compared with the logarithmic form of Doolittle's equation (4.9), namely:

Equation 4.18
$$\ln \eta = A' + \frac{BV_r}{V_f}$$

shows that $\ln \eta$ (and not η itself) is inversely proportional to V_f if the constant A' is neglected.

If $V_f = 0$—which is the case at T_r—then the viscosity *does* become infinity no matter whether the *modified* Batschinski equation (4.17) or Doolittle's equation is used. If creep is nevertheless observed in materials below T_r, then this must be due to the compressibility of the units themselves which may or may not be recoverable depending on the balance of viscous and elastic responses within the internal microcosm of the units.

Bridgman, who has made an exhaustive study of the effects of pressure on the viscosity of liquids, has come to the conclusion that the Batschinski equation is quantitatively incorrect at both constant pressure and at constant volume; but it cannot be denied that there is a close connection between viscosity and free volume and in that sense Batschinski's work strongly supports the mechanism of flow just presented.

79

The free volume concept can be applied to permeability in the same way as to viscosity. If flow is a rate process, so is permeability, since diffusion plays a dominant part in both. If the rate of diffusion of various gases through polymer films is determined as a function of temperature it becomes apparent that, in general, the rate of diffusion decreases with decreasing temperature. This is precisely what one would expect since the diffusing molecules will find it increasingly difficult to penetrate a polymer mass as the free volume of that mass decreases. If the statistical mean capillary cross-sectional area of the free volume 'channels' in the polymer mass becomes less than the diameter of the penetrating molecule then permeation will effectively cease.* For all gases the permeability must be at a minimum at T_r and show no further decrease below that temperature. Indeed the minimum permeability for any given gas will be reached somewhat above T_r and will be further and further removed from T_r as its molecular diameter increases. Thus, it should be possible to determine the statistical mean capillary cross-sectional area of the free volume channels at any given temperature by finding a gas of known molecular dimensions which just reaches its minimum permeability with respect to the polymer mass at that temperature.

According to Graham's law of diffusion, the rate of the process follows an inverse square root law with respect to density. This implies not only that the experimental conditions are fixed with respect to temperature and pressure when making measurements on different gases—it also implies that the cross-sectional area of the orifice through which the gases are diffusing is constant. Temperature changes will have the same effect on the gases, no matter what type of orifice they will diffuse through. But whilst a *glass* or *metal* orifice is not greatly affected by temperature changes, the diffusion of a gas through a *polymer film* will be greatly influenced by the highly temperature-dependent 'orifice' sizes, that is to say the statistical mean cross-sectional area of the free volume channels of the polymer, which is an inverse function of its density.

The effect of pressure under ambient conditions is negligible. But at pressures above about 200 psi the compression of molecules gives rise to an exponential increase in viscosity similar to that of the temperature effect. Dow, Dibert and Fink [7] have reported this type of behaviour for mineral oils up to pressures of 4,000 atm. Bridgman [8] has found that the viscosity of silicone fluids increases by 10^7 poise under an applied pressure of 10,000 atm. In this pressure range the viscosity

---

* This ignores possible 'tunnelling' effects due to adsorption/desorption or solution mechanisms.

increase is *more* rapid than exponential and it is considered that at about 30,000 atmospheres silicone fluids become effectively glass-like.

It is not difficult to reason out why an increase in pressure should cause viscosity increases comparable to those obtained by cooling liquids. In both cases the free volume is decreased. Furthermore, the energy available to flow units for surmounting the activation energy hump is less at reduced temperatures, whilst at increased pressures the flow units are placed into a tight environmental strait-jacket which raises the energy hump against positional change as shown in Fig. 4.5 below:

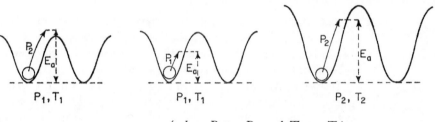

P_1, T_1 P_1, T_1 P_2, T_2

(where $P_2 > P_1$ and $T_2 > T_1$)

Fig. 4.5

There should therefore exist a pressure P which, when applied to a material, will cause a similar viscosity increase at constant temperature as a temperature decrease would cause without additionally applied pressure.

Since liquids are generally not very compressible, the pressures which are required to reduce the free volume significantly are of a very high order indeed. Carley has reviewed the effect of static pressure on polymer melt viscosities [9]. He pointed out that with pressures of the order encountered at injection nozzle and extrusion die entrances the viscosity of polymer melts is very noticeably increased. The viscosity/pressure function is not a simple one—it is certainly exponential. Bridgman in his work on liquids also noted that the pressure effect on viscosity increases as the molecular structure becomes more complex. The complexity of polymer melts is so great that one would expect a correspondingly important effect. In order to evaluate this, Maxwell and Westover constructed double-piston rheometers so that the static pressure could be regulated on either side. The practical significance of the fact that viscosity increases with increasing pressure lies in this: The more pressure is applied to force melt through an orifice, the more 'it does not come out.' Or, as Maxwell puts it—rather more elegantly—' the application of higher and higher pressures becomes, in effect, self defeating with no nett advantage to the moulder.' The effect is more pronounced with polystyrene than with polythene, as both Maxwell and Westover found:

MELT VISCOSITY INCREASE WITH PRESSURE (PSI)

		from (psi)	*to (psi)*	*Viscosity increase over the range*
Maxwell	Polythene	8,000	26,000	× 5·7
	Polystyrene	8,000	26,000	×135
Westover	Polythene	2,000	25,000	× 5·6
	Polystyrene	2,000	25,000	×135

The difference between polythene and polystyrene is not unexpected. In the former the flow unit is as simple as it can be, in the latter the pendant phenyl groups make the flow unit substantially more cumbrous and complex.

If the pressures applied are so large that liquids become glassy solids and eventually (theoretically) solids without free volume, then on further increasing the pressure the internal microcosm of the flow units will begin to be compressed against molecular and atomic forces, causing a complete collapse of the structure of matter and resulting in a dense mass of packed subatomic particles. The pressures required to cause this ultimate collapse are, of course, unimaginably large and the resulting mass would have an enormous density. Such pressures cannot be experimentally realised, but it is thought that gravitational fields which exert forces of this order might exist in the universe.

We are really rather limited in our scope of changing free volume by applying such pressures as are available to us; a change in free volume is much more readily effected by changing the *temperature*. The coefficient of thermal expansion of plastics (both in the liquid and solid state) is high and small changes in temperature can therefore affect the viscosity of a polymer quite markedly. As for the two causes of decreasing viscosity (the diminishing activation energy of the actual jumping process and the decrease in free volume), the latter seems to be the more important. The activation energy of the jumping process appears to be quite small and most of the increase in viscosity with falling temperature would appear to be due to the contraction of free volume in the material. It is, however, common to lump the two effects together as the overall activation energy of viscous flow without trying to separate it into (a) the jump energy and (b) the energy necessary for the creation of new 'holes' in the structure.

The effect of shear rate

A highly significant development concerning the effect of shear rate on polymer melts has been occasioned by a paper by Van der Vegt and Smit [10] read at the Conference on Advances in Polymer Science and Technology, London, September, 1966 which was jointly sponsored by the Plastics and Polymer Group of the Society of the Chemical Industry, the Institution of the Rubber Industry and the Plastics Institute (see also [11]). Van der Vegt and Smit studied the melt viscosity of a number of thermoplastics and unvulcanised elastomers in a capillary rheometer. They noticed that under certain conditions the apparent viscosity sharply increased with shear rate and that this increase might be so great as to completely inhibit flow. This effect, first observed by Van der Vegt and Smit with polypropylene, polythene, cis-1,4 polybutadiene and cis-1,4 polyisoprene, was also observed by other workers with PVC and polystyrene. The suggestion was made that the viscosity increase is due to orientation-induced crystallisation in the melt and the striking feature of this concept is that this explanation can be applied equally to polymers which are normally considered to be amorphous *as well as* to those possessing substantial crystallinity at sub-melt temperatures. The assumption was confirmed by the X-ray diffraction patterns of the contents of the capillary after quenching and it was clearly seen that a highly oriented crystalline structure is present during high shear rate flow, whilst at lower shear rates the crystallisation is random and isotropic. Crystallisation in the melt is comparable to orientation crystallisation of a highly and rapidly stretched rubber vulcanisate. This effect is unquestionably unconnected with the pressures necessary to produce those high shear rates, since these are quite moderate (of the order of 100 kg/cm^2 at 165°C in polypropylene) and since far greater pressures are required to induce genuine pressure-induced viscosity increases of the Bridgman type. However, in extrusion and especially in moulding processes the shear-rate-induced viscosity increase through melt crystallinity and the pressure-induced viscosity increase through free volume reduction are undoubtedly both present. It may well be that, now that the existence of the former has been clearly shown to be possible in all types of thermoplastics, it will also be proved that shear-rate-induced viscosity increases are quantitatively more important than those induced by *direct* pressure on the melt.

The effect of molecular weight

As the molecular weight of a polymer increases, so the melt viscosity increases because the activation energy of the process increases with increasing mass of the flow unit. But once the molecular weight has reached a certain value, the chains will begin to entangle and whole chains can no longer be the flow units. This is proved by the fact that on increasing the molecular weight further, the activation energy of flow remains virtually constant. Any further increases in molecular weight only cause linear increases instead of exponential increases in viscosity. That is to say, the proportionality constant A in the Arrhenius equation becomes a function of viscosity and therefore of molecular weight. The empirical relationship

Equation 4.19
$$A = \text{const.} \times M^{3.4}$$

applies to many polymers, both linear and branched, and in concentrated solution as well as in the melt.

The unit of flow is believed to be a chain segment of about 50 aliphatic carbon atoms and the mass involved accounts for the fact that T_r in polymers is between 200 and 400°C above the absolute zero. By determining T_r for polymers of sufficiently high molecular weight to form flow units of constant length it should therefore be possible to determine the mass of flow units in chain structures relative to the mass of 50 methylene groups.

To put it another way, in homo-carbon chain structures of sufficiently great chain length

$$\frac{L_B}{L_A} = \frac{(T_r)_A}{(T_r)_B} \frac{M_B}{M_A},$$

where

L_A and L_B = length of flow unit (number of carbon atoms) in aliphatic vinyl type polymer chains A and B respectively,

$(T_r)_A$ and $(T_r)_B$ are the reference points in °K for molecular relaxation processes of polymers A and B respectively, and

M_A and M_B are the molecular weights of the flow units in the two polymers.

We know that the flow unit in linear polythene is about 50 carbon atoms, or 25 repeating units, so that the molecular weight of the flow unit is about 700. The values of T_r can be obtained from viscometric data in the melts by curve fitting using the modified Arrhenius equation as shown by Miller (see above), while the molecular weight of the repeating units is known. Using linear polythene as the reference material we can therefore write:

$$L_B = \frac{(T_r)_B}{M_B} \cdot \frac{700}{(T_r)_A}$$

where $(T_r)_A$ is the T_r in °K of linear polythene.

Conversely, if L_B is known, then $(T_r)_B$ can be calculated from

$$(T_r)_B = \frac{(T_r)_A M_B L_B}{700}$$

It will be interesting to see how $(T_r)_B$, calculated from independently determined values of L_B, compares with experimentally determined values of $(T_r)_B$.

The first group of homo-carbon-chain aliphatic polymers which suggests itself for study would comprise the following:

> Linear Polythene (polymethylene)
> Polypropylene
> Poly (4 methyl pentene-1)
> Polystyrene and polychlorostyrenes
> Polyacrylates
> Polymethacrylates
> Polyacrylonitrile
> Polyvinyl chloride
> Polyvinylidene chloride
> 'Penton'

Modifications will probably have to be made for polymers with bulky branches such as phenyl groups, for hetero-chain polymers and for polymers containing linear sequences of aromatic, cycloaromatic and heterocyclic ring systems, since the flow units in those polymers may differ substantially from the linear polythene value of about 50 carbon atoms. Modifications will also be necessary when linear polymers with exceptionally high chain mobilities are considered. Thus, it has been shown as long ago as 1945 [12] that the unit of relaxation of polychloroprene could be as low as 17 carbon atoms even in a high molecular weight polymer. Naturally, any plasticising additives and solvents will also have the effect of reducing the size of the flow unit since smaller regions will be able to move independently as a result of the loosening of the bulk structure. This, together with equivalent changes in mobility as a function of temperature has been intensively studied especially by mechanical and electrical methods, although dynamic techniques have proved much more powerful than the static ones to which we have mainly confined ourselves so far.

The molecular interpretation of physical relaxation processes in terms of structural rearrangement has received a great deal of attention from Boltzmann onward and the whole literature on diffusion processes and energy barriers is relevant. The reader is referred to a paper by Müller which includes twenty-four references [13] and to Treloar's admirable monograph [16].

Temperature viscosity relationships in non-Newtonians

It is now appropriate to extend the temperature/viscosity relationship of liquids to non-Newtonian power-law liquids. We have seen that in Newtonians the viscosity of a given polymer is a function of temperature only, if the molecular weight remains constant. This is also true for non-Newtonians, but in the latter the viscosity additionally depends on the shear rate $\dot{\gamma}$ or shear stress τ and it becomes necessary to keep either $\dot{\gamma}$ or τ constant in order to isolate the temperature effect. In general, the viscosity change with temperature at constant $\dot{\gamma}$ is different from the viscosity change with temperature at constant τ. This can be readily proved mathematically [14]: The total differential of η as $f(\tau, T)$ is:

$$d\eta = \left(\frac{\partial \eta}{\partial T}\right)_\tau dT + \left(\frac{\partial \eta}{\partial \tau}\right)_T d\tau$$

If $\dot{\gamma}$ is constant, this becomes:

$$d\eta = \left(\frac{\partial \eta}{\partial T}\right)_\tau dT + \left(\frac{\partial \eta}{\partial \tau}\right)_T \left(\frac{\partial \tau}{\partial T}\right)_{\dot{\gamma}} dT$$

or

$$\left(\frac{\partial \eta}{\partial T}\right)_{\dot{\gamma}} = \left(\frac{\partial \eta}{\partial T}\right)_\tau + \left(\frac{\partial \eta}{\partial \tau}\right)_T \left(\frac{\partial \tau}{\partial T}\right)_{\dot{\gamma}}$$

Rearranging:

$$1 = \frac{\left(\dfrac{\partial \eta}{\partial T}\right)_\tau}{\left(\dfrac{\partial \eta}{\partial T}\right)_{\dot{\gamma}}} + \frac{\left(\dfrac{\partial \eta}{\partial \tau}\right)_T \left(\dfrac{\partial \tau}{\partial T}\right)_{\dot{\gamma}}}{\left(\dfrac{\partial \eta}{\partial T}\right)_{\dot{\gamma}}}$$

or

$$\frac{\left(\dfrac{\partial \eta}{\partial T}\right)_\tau}{\left(\dfrac{\partial \eta}{\partial T}\right)_{\dot{\gamma}}} = 1 - \left(\frac{\partial \eta}{\partial \tau}\right)_T = 1 - \left(\frac{\partial \eta}{\partial \tau}\right)_T \left(\frac{\partial \tau}{\partial \eta}\right)_{\dot{\gamma}}$$

Differentiation of the basic equation $\tau = \eta\dot{\gamma}$ with respect to η at constant $\dot{\gamma}$ gives:

$$\left(\frac{\partial\tau}{\partial\eta}\right)_{\dot{\gamma}} = \dot{\gamma}$$

and this reduces the above equation to:

Equation 4.20

$$\frac{\left(\dfrac{\partial\eta}{\partial T}\right)_{\tau}}{\left(\dfrac{\partial\eta}{\partial T}\right)_{\dot{\gamma}}} = 1 - \dot{\gamma}\left(\frac{\partial\eta}{\partial\tau}\right)_{T}$$

In pseudoplastics η reduces as τ increases, therefore $(\partial\eta/\partial\tau)_T$ is −ve and the right-hand side of Equation 4.20 must be >1. In dilatants η increases as τ increases, therefore $(\partial\eta/\partial\tau)_T$ is +ve and the right-hand side of 4.20 must be <1.

Hence, in the general case,

$$\left(\frac{\partial\eta}{\partial T}\right)_{\tau}\bigg/\left(\frac{\partial\eta}{\partial T}\right)_{\dot{\gamma}} \neq 1, \text{ q.e.d.}$$

In general, then, for a pseudoplastic $(\partial\eta/\partial T)_{\tau} > (\partial\eta/\partial T)_{\dot{\gamma}}$ and only when the second term on the right-hand side of (42.0) is zero do the two η derivatives become equal. This is the case when *either* $(\partial\eta/\partial\tau)_T = 0$ (the definition of a Newtonian) *or* when $\dot{\gamma} = 0$. This raises the interesting fact that *all fluids at zero shear rate* are in fact Newtonians. In practice Newtonian behaviour is observed for $\dot{\gamma}$ *close to* zero and it is of interest to determine at what $\dot{\gamma}$ the fluid begins to deviate from Newtonian behaviour.

We can express the T-dependence of η in one of two ways:

Equation 4.21
$$\eta = A\,e^{-E\dot{\gamma}/RT} \quad \text{and} \quad \eta = A\,e^{-E\tau/RT}$$

At very low shear stresses or shear rates (τ or $\dot{\gamma} \to 0$, see above) it was found that $E_{\dot{\gamma}} \equiv E_{\tau}$. This can be derived from Equation 4.20 by introducing the partial derivatives of (4.21),

Equation 4.22
$$\left(\frac{\partial\eta}{\partial T}\right)_{\tau} = -\eta\frac{E_{\tau}}{RT^2} \quad \text{and} \quad \left(\frac{\partial\eta}{\partial T}\right)_{\dot{\gamma}} = -\eta\frac{E_{\dot{\gamma}}}{RT^2}$$

so that:

Equation 4.23
$$\frac{E_{\tau}}{E_{\dot{\gamma}}} = 1 - \dot{\gamma}\left(\frac{\partial\eta}{\partial\tau}\right)_{T}$$

In power law fluids E_{τ} and $E_{\dot{\gamma}}$ are not independent, but related by the simple expression

8

$$\frac{E_{\dot{\gamma}}}{E_{\tau}} = n$$

where n is the *power index* which denotes the degree of deviation from the Newtonian behaviour characterised by a power index of unity.

Hence $n < 1$ for a pseudoplastic and $n > 1$ for a dilatant. The determination of activation energies thus affords a means of calculating this important parameter.

An example of how this may be put to good use is given below. It is taken from McKelvey's *Polymer Processing*, with additional detail of the numerical computation:

How can the same change in temperature ΔT produce different viscosity changes in a non-Newtonian liquid?

Consider a non-Newtonian liquid in the space between the cone and plate of a rotational viscometer the cone of which is driven by a *constant speed* device. On increasing the temperature the viscosity will drop and less torque will be required to drive the cone. The shear rate $\dot{\gamma}$ is constant and the viscosity change is given by:

Equation 4.24
$$d\eta = \left(\frac{\partial \eta}{\partial T}\right)_{\dot{\gamma}} \partial T$$

which, because of Equation 4.22 can be written:

Equation 4.25
$$d\eta = -\eta \frac{E_{\dot{\gamma}}}{RT^2} \, dT \quad \text{or} \quad \ln \eta = -\frac{E_{\dot{\gamma}}}{RT^2} \, dT$$

which on integration between the limits of T_1 and T_2 gives:

Equation 4.26
$$\ln \frac{\eta_2}{\eta_1} = -\frac{E_{\dot{\gamma}} \Delta T}{R T_1 T_2}$$

The negative sign on the right-hand side of Equation 4.26 indicates that the viscosity decreases as the temperature increases.

Supposing on the other hand, that the drive of the cone is a *constant torque* device; we would now obtain a different result. The shear stress τ will now be constant and the shear rate $\dot{\gamma}$ will therefore increase as the temperature is increased. It is therefore possible to use the same argument as that employed in the derivation of Equation 4.26 and the analogous result affords Equation 4.27:

Equation 4.27
$$\ln \frac{\eta_2}{\eta_1} = -\frac{E_{\tau} \Delta T}{R T_1 T_2}$$

Since $E_{\tau} > E_{\dot{\gamma}}$, the viscosity decrease at constant shear rate is *less* than the viscosity decrease at constant shear stress.

Consider the following problem:

ACTIVATION ENERGIES FOR POLYTHENE BETWEEN 108 AND 230°C [15]

$\dot{\gamma}$ (sec^{-1})	$E_{\dot{\gamma}}$ (kcal mole^{-1})	τ (dynes cm^{-2})	E_{τ} (kcal mole^{-1})
0	12·8	0	12·8
10^{-1}	11·4	10^4	15·0
10^0	10·3	10^5	17·8
10^1	8·5	10^6	19·0
10^2	7·2		
10^3	6·1		

At a temperature of 174°C and a standard shear rate of $\dot{\gamma} = 1$ sec^{-1}, the viscosity of a polythene (whose activation energies had been determined by Philippoff and Gaskins [15] and which are given above) was found to be 31,500 poise. Assuming the power law to be applicable over the range of temperatures and viscosities involved, calculate the viscosity of the polythene at 230°C and at a shear rate of 100 sec^{-1}.

The power law $\tau = \eta \dot{\gamma}^n$ is used in the form

$$\eta = \eta^0 \left(\frac{\dot{\gamma}}{\dot{\gamma}^0}\right)^{n-1} = \eta^0 \left(\frac{\tau}{\tau^0}\right)^{(n-1)/n}$$

where superscript o refers to an arbitrary standard state. If the arbitrary standard state is taken as $\dot{\gamma}^0 = 1$ sec^{-1}, then η^0 at 230°C can be obtained as follows:

Taking logarithms of the preceding equation we get:

Equation 4.28
$$\ln \eta = \ln \eta^0 + (n - 1) \ln \frac{\dot{\gamma}}{\dot{\gamma}^0}$$

On differentiating Equation 4.28 the second right-hand term disappears since it is a constant and we obtain:

Equation 4.29
$$\left(\frac{\partial \ln \eta^0}{\partial T}\right)_{\dot{\gamma}} = \left(\frac{\partial \ln \eta}{\partial T}\right)_{\dot{\gamma}}$$

Differentiating the logarithmic form of Equation 4.25, we obtain

$$\frac{\partial \ln \eta}{\partial T} = -\frac{E_{\dot{\gamma}}}{RT^2}$$

which because of Equation 4.29 becomes

$$\frac{\partial \ln \eta^0}{\partial T} = -\frac{E_{\dot{\gamma}}}{RT^2}$$

This, on integration between the limits of T_1 and T_2 yields:

89

Equation 4.30
$$\ln \frac{\eta_2{}^0}{\eta_1{}^0} = \frac{E_{\dot\gamma}}{R}\left(\frac{1}{T_2} - \frac{1}{T_1}\right) = \frac{E_{\dot\gamma}}{R}\frac{\Delta T}{T_1 T_2}$$

Taking the value of $E_{\dot\gamma}$ from Philippoff and Gaskins's table as 10·3 kcal mole^{-1} and $R = 1\cdot98$ cals

$$\log \frac{\eta_2{}^0}{\eta_1{}^0} = \frac{10\cdot3 \times 10^3}{2\cdot303 \times 1\cdot98}\left(\frac{1}{230 + 273} - \frac{1}{174 + 273}\right)$$

$$= 2\cdot26 \times 10^3\left(\frac{1}{503} - \frac{1}{447}\right)$$

$$= 2\cdot26 \times 10^3(0\cdot001987 - 0\cdot002235)$$

$$= -2\cdot26 \times 0\cdot248 = -0\cdot56$$

Since $\eta_1{}^0 = 31{,}500$ poise,

$$\log \eta_2{}^0 = \log 31{,}500 - 0\cdot56 = 4\cdot4983 - 0\cdot56 = 3\cdot9383$$

and

$$\eta_2{}^0 = 8{,}680, \text{ say } 8{,}700 \text{ poise}$$

This, then is the standard viscosity ($\dot\gamma^0 = 1$ sec^{-1}) at $T_2(230°\text{C})$. To find the viscosity η_2 at T_2 and 100 sec^{-1} we apply the power law in the form

Equation 4.31
$$\eta_2 = \eta_2{}^0 \dot\gamma^{\bar{n}-1}$$

where $\bar{n}$ is the average of the power index n between the shear rates $\dot\gamma$ of 1 and 100 reciprocal seconds. We do not know n, the power index at $\dot\gamma = 1$ sec^{-1}, but we can obtain it readily enough from the ratio of activation energies ($n = (E_{\dot\gamma}/E_\tau)$). But before we can do this we must know (or at least make an inspired guess at) the order of magnitude of the shear stress τ at the standard shear rate $\dot\gamma = 1$ sec^{-1}. We can make such a guess by saying that the *order* or magnitude of the viscosity will be the same, irrespective of whether the temperature is 174°C—we found the viscosity to be 8,700 poise, say, of the order of 10^4 poise, and since

$$\eta^0 \simeq \frac{\tau}{\gamma^0}$$

(only very roughly, since $n \neq 1$), then

$$10^4 \simeq \frac{\tau}{1}$$

so that the shear stress τ will also be of the order of 10^4 dynes cm^{-2}.

Looking up the activation energy E_τ opposite $\tau = 10^4$ in Philippoff and Gaskins's table, we find it to be 15·0 and since $\dot\gamma^0 = 10^0$, $E_{\dot\gamma}$ (from the same table) is 10·3.

The power index n, at $\dot\gamma = 1$ sec^{-1} is therefore:

$$n_1 = \frac{E_{\dot\gamma}}{E_\tau} = \frac{10·3}{15·0} = 0·68$$

We next want to know the power index n_2 at $\dot\gamma = 100$ sec^{-1}. This we can look up directly in the table where we find it to be 7·2. Again we shall have to make a guess at τ in this state and we assume that it is of the order of 10^5 dynes cm^{-2} at $\dot\gamma = 100$ sec^{-1}, on the same grounds on which we assumed that it was of the order of 10^4 dynes cm^{-2} at $\dot\gamma = 1$ sec^{-1} before.

Looking up E_τ opposite $\tau = 10^5$ in Philippoff and Gaskins's table we find it to be 17·8, whence

$$n_2 = \frac{E_{\dot\gamma}}{E_\tau} = \frac{7·2}{17·8} = 0·4$$

The mean power index is therefore

$$\bar n = \frac{n_1 + n_2}{2} = \frac{0·68 + 0·4}{2} = 0·54$$

and introducing this value into Equation 4.31

$$\eta_2 = \eta_2{}^0 \dot\gamma^{\bar n - 1}$$

we get:

$$\eta_2 = 8{,}700 \times (100)^{-0·46} = 1{,}040 \text{ poise}$$

Checking back roughly,

$$\left(\eta_2 \cong \frac{\tau}{\dot\gamma} \right)$$

we find that

$$1{,}040 \cong \frac{\tau}{100} \quad \text{or} \quad \tau \cong 104{,}000 \text{ dynes cm}^{-2}$$

that is to say, it is indeed of the order of 10^5 and the guess was a good one.

If we had been an order of magnitude out one our guess—say $\tau \cong 10^6$ dynes cm^{-2}: then E_τ, according to the table, would be 19·0, n_2 would be 7·2/19·0 = 0·38 and $\bar n$ would be

$$\frac{0·68 + 0·38}{2} = 0·53$$

giving η_2 as 1,000 poise.

Had the guess been $\tau \cong 10^4$ dynes cm^{-2}, then E_τ would be

15·0, n_2 would be $7·2/15·0 = 0·48$ and $\bar{n}$ would be

$$\frac{0·68 + 0·48}{2} = 0·58$$

giving η_2 as 1,250 poise.

If we therefore happen to be wrong in guess by even two orders of magnitude, the answer would still be a viscosity of the order of 10^3 poise and checking back using:

$$\eta \cong \frac{\tau}{\dot{\gamma}}$$

we find that

$$10^3 \cong \frac{\tau}{10^2} \quad \text{or} \quad \tau \cong 10^5$$

It is then possible, after making a poor guess, to correct this by looking up E_τ at what is now known to be the correct order of magnitude of τ, namely 10^5, and obtaining the correct value for n_2 and eventually the correct value (1,040 poise) for η_2 at 230°C and 100 sec^{-1}.

Determination of activation energies

Activation energies are obtained in the following way:

(i) Plot τ vs $\dot{\gamma}$ over a range of shear rates at various temperatures ($T_5 > T_4 > T_3 > T_2 > T_1$) to obtain a family of curves as shown in Fig. 4.6 below.

(ii) Draw a line Y representing some constant shear rate $\dot{\gamma}$ across the curves and determine the slopes $\tau/\dot{\gamma}$ of the tangents at the points of intersection (dotted in Fig. 4.6). These slopes are the viscosities at constant shear rate. Using the Arrhenius equation

$$\ln \eta = \text{const.} - \frac{E_{\dot{\gamma}}}{RT},$$

plot $\ln \eta$ vs $1/T$, when a straight line is obtained with slope $E_{\dot{\gamma}}/R$ and $E_{\dot{\gamma}}$ follows.

(iii) Repeat the procedure at a number of other shear rates and tabulate the values of $E_{\dot{\gamma}}$ corresponding to each value of $\dot{\gamma}$.

(iv) Draw a line X representing some constant shear stress τ across the curves and determine the slopes $\tau/\dot{\gamma}$ of the tangents at the points of intersection (again, dotted in Fig. 4.6). These

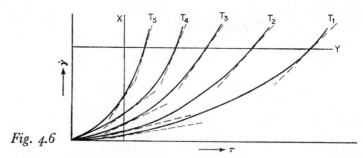

Fig. 4.6

slopes will now represent the viscosities at constant shear stress. Using the Arrhenius equation

$$\ln \eta = \text{const.} - \frac{E_\tau}{RT}$$

determine E_τ from the resulting linear plot of $\ln \eta$ vs $1/T$ from the slope E_τ/R.

(v) Repeat the procedure of (iv) at a number of other shear stresses and tabulate the values of E_τ corresponding to each value of τ.

This method shows how much of the rheological information on a material is ultimately dependent on shear rate/shear stress curves in the liquid state. In the solid state the 'static' stress/strain curves fulfil an analogous function and this will be the subject of Chapter 6.

5 Melt fracture and melt elasticity

Flow disturbance in extrusion—suggested causes based on the concepts of relaxation phenomena, free volume, molecular architecture and departure from laminar flow. General discussion of machine and material variables from the point of view of processing and product quality. Chain entanglement as a cause of the manifestation of melt elasticity effects (die swell); critical chain length; effects of polymer history on die swell. Return to molecular interpretation. Melt instability explained, as a result of very recent work, as an inevitable consequence of an underlying and intrinsic dynamic situation caused by shear acceleration at the die entrance.

THE types of flow which one would normally expect to occur in polymer melts have now been discussed. Attention has been focused on power-law liquids, with occasional reference to the special Newtonian case when the power index is equal to unity in order to check conclusions qualitatively.

It is now desirable to look at some of the implications from the point of view of polymer processing.

Merz, Kircher and Hamilton [1] have shown that the shear rate of plasticised PVC increases overall by about a decade in each step from compression moulding through milling and calendering through extrusion to injection moulding, whilst the maximum shear stress at the wall increases at a rather lesser rate. Of the processes mentioned extrusion is the most readily amenable to flow analysis, since the boundaries are well defined (unlike in calendering) and since the process operates under continuous steady state conditions (unlike injection moulding) in which the variables can be fairly closely controlled.

Extrusion has therefore been the process on which a great deal of experimental work has been published.

It has long been known that grossly irregular extrudate is invariably obtained when excessive pressures are used, especially when an insufficiently high melt temperature raises the melt viscosity to such an extent that very high pressures must be applied to obtain any extrudate at all at a reasonable rate. The gross irregularities suggest that the phenomenon is caused by flow disturbance which produces 'melt fracture' at a distance so close to the point of extrudate emergence that the dwell time in the die channel is insufficient for laminar flow to establish itself. The longer the die, the greater the probability that the memory of the disturbance will have worn off and that the melt fracture will have mended itself. The post-die appearance of melt fracture is therefore dependent on the dwell time in the die channel. This dwell time, in turn, is a function of the applied pressure and of the die geometry. In order to minimise the flow disturbance which occurs in the first place one would have to identify the exact locus where it arises. Tordella [2 and 3] has pointed out some time ago that the obvious site of melt fracture is the die inlet region. Kendall [4] proved this assertion by altering the inlet geometry from a square cut to tapers of varying angles and showed that for any given set of operating conditions the degree of gross extrudate distortion was a function of the die inlet taper angle. The gentler the taper the smaller will be the elastic stresses and the better able will the emerging stream of melt be to sustain them. If the rate of throughput is known then the known die geometry enables one to calculate the dwell time of melt in the channel proper and by noting the varying pressure the dwell time can be determined which marks the transition from even to distorted extrudate. This time will vary with melt viscosity and therefore temperature, but at any one temperature it is a constant value characteristic for the relaxation of the flow units. To be more precise, however, it should be pointed out that, whilst for any given polymer of sufficiently high molecular weight the flow units are essentially similar in length, they do cover a range around an average length, so that the relaxation time also represents only an average value. The more polydisperse a material is and, in particular, the greater the proportion of low molecular weight material, the wider will be the distribution of relaxation times. High molecular weight fractions will not affect the relaxation time to the same degree since the flow unit is virtually constant once the molecular weight is of a sufficiently high order. The relaxation time is also a function of free volume and is therefore very susceptible to temperature changes. It is for this reason that the die tip is usually heated to a temperature a few degrees higher than the rest of the die. Plasticiser also increases the free volume and reduces relaxation times.

In order to make a good profile at a fast rate it is therefore desirable to use low molecular weight polymers. On the other hand, low molecular weight material poses take-off handling problems since the extrudate emerges at relatively low viscosity; moreover, it is desirable to use the highest possible average molecular weight containing the least possible amount of low molecular weight fractions in order to achieve the highest possible mechanical strength. Inevitably, these conflicting requirements demand a compromise.

In the special case of wire covering, however, where thermal degradation of the insulating plastic at long dwell times causes a deterioration in electrical properties, where the production rates must be very high for obvious economic reasons and where mechanical strength is of secondary importance, polymers of quite low average molecular weight can and are in fact being used. In wire-coating dies (which are crosshead dies) flow disturbance is very prone to occur and this, again, suggests the use of low molecular weight material. Similar considerations apply to fibre and monofilament extrusion. Here the die diameters are small and must be short in order to minimise pressure losses, the production rates must be high and the inlets may be square cut. The mechanical strength of the extrudate does not necessarily have to be great at this stage since the required ultimate strength can be procured by subsequent cold drawing and orientation which, incidentally, will also virtually eliminate extrudate distortion because of the high draw-down ratio.

In tubular film melt fracture is not normally encountered because the shear stresses are not such as to exceed the critical value for gross flow disturbance. Kendall did, however, introduce turbulence by using rather unusual extrusion settings. He also introduced dyed marker masterbatch material between the screw end and the die entrance and noted that the oscillatory movement of the marker line was out of phase with that of the undrawn extrudate itself. This behaviour is typical of elastic disturbance. When the disturbance was sufficiently great, then fracture of the marker line occurred. In practice, melt distortion (especially in high molecular weight polythenes) can be avoided by:

1. Reducing the extrusion rate, or

2. Using a die of large cross section and increasing the draw-down to compensate for the increase in cross section. Either or both these measures will reduce the shear stress below the critical point at which gross flow disturbance occurs.

In chill roll casting of film from a slit die there is little danger of die entrance flow disturbance because

1. Polymers of comparatively low molecular weight are used, and

2. Extrusion temperatures are high and the viscosity is correspondingly low. The situation is similar to monofilament extrusion, inasmuch as in both a high draw-down ratio is employed.

In extrusion blow moulding flow disturbance is very prone to occur because low melt temperatures and high shear stresses are encountered with the high molecular weight material which is normally used in order to obtain the highest possible mechanical strength. The resulting melt fracture effects are visually confined to the inside of bottles because the mould walls smooth out the external irregularities on blowing—but the internal irregularities of the blown parison are visible in unpigmented bottles of clear or translucent material. For this reason bottle blowing dies usually have fairly long die lands.

In injection moulding the material requirements, the requirements of mould design, the considerations for optimum operating conditions and the requirements of product quality are highly conflicting. A low average molecular weight material is desirable for the following (and other) reasons:

1. Low viscosities facilitate mould filling;

2. The operation can be carried out at lower temperatures and so reduce the risk of thermal degradation—or conversely, if the temperature is not reduced, one obtains the benefit of lower pressure requirements for mould filling and clamping;

3. Weld lines and weaknesses in the moulding arising from weld lines are minimised;

4. Intricate mouldings, possibly even involving inserts, can be more readily achieved.

On the other hand, high average-molecular-weight material is desirable for the following (and other) reasons:

1. Greater mechanical strength of the moulding;

2. Faster cycle times are possible because the difference in temperature between the melt in the barrel and the mould is great and the moulding can therefore be ejected after a shorter mould-closed period, although this will also cause residual stresses in the moulding which may result in distortion as strains are relieved.

Again, it is clear that a compromise is necessary. A turbulent flow into the mould is not undesirable since it ensures that random orientation of the flow units is promoted and that the mechanical properties of the moulding are reasonably isotropic. In any case, with the channel restrictions, with the high pressures involved compared to extrusion and with the high rates of injection (high shear rates) necessary to ensure complete mould filling before the melt begins to freeze it is quite impossible to avoid turbulence anyway.

We have now considered the phenomenon of melt fracture with special reference to extrusion processes. We have seen how it is caused and how it can be avoided or minimised. After agreeing, however, that melt fracture is due to flow disturbance at the die entrance, it must also be said that there may be other sites where melt fracture of a secondary kind may be produced. This secondary melt fracture does not result in gross distortion but causes more or less severe surface effects known as 'shark skin' and 'orange peel'.

E. T. Severs observed that at low shear stress a rod of extrudate could be snipped off cleanly at the die orifice—indicating good adhesion of the melt to the die wall. But as the shear stress is raised above some critical value the entire contents of the die could be withdrawn on giving the extrudate a sudden vigorous pull. This indicates poor adhesion of the melt to the die wall. V. G. Kendall [11] considers that shark skin and orange peel are surface defects arising at the die exit due to differential recovery between the highly strained extrudate skin and the less strained material in the body of the melt to which slip-stick effects at the wall may contribute, pointing out that these defects can occur at quite moderate flow rates at which inlet melt fracture does not manifest itself. A similar view involving slip-stick effects was also put forward by Tordella [5] who, however, could not find any evidence that the initiation of roughness was due to differential elastic recovery between the various portions of the emerging polymer stream. Tordella showed that above some critical stress the extrudate had the appearance of a string of sausages. This was no doubt due to primary or inlet melt fracture. But at *slightly* lower stresses a fine roughness is observed. This may also be present at higher stresses, superimposed on to the sausages and suggests that minor relaxation phenomena occur at the die wall. When the stress becomes large enough to make its dissipation in 'multimicro' steps impossible, then periodic major discontinuities will occur either instead of or additionally to the surface roughness.

The instability initiated at the die wall results in fine scale roughness, then in slip flow and eventually in gross distortion and arises from elastic failure of the extrudate at certain critical elastic strains. Tordella proved this by capillary and rotational viscometry and flow birefringence experiments.

The elastic (recoverable) strain present in viscoelastic polymer melts shows itself clearly in the die swell of the emerging polymer stream. The swelling ratio, defined as the ratio between the diameters of the melt profile and the die (as produced, for example, in melt flow index determinations), gives some indication of the elastic properties. Mieras [5a] has shown that good quantitative agreement in the assessment of the elastic properties of polypropylene melts can be obtained between experimental techniques based on the measurement of die swell and on melt birefringence measurements. The swelling is independent of the die length at low output rates, but at high output rates it is less pronounced. This has been explained in various ways [6]:

1. Re-randomisation of polymer chains which have become flow-orientated whilst passing through the die.

2. Transition from a near-parabolic velocity distribution at the die exit to a constant velocity of the extrudate some way down the drawn rod. The elasticity of the highly strained skin slows down the core velocity, so that the cross sectional area must increase. The converse possibility of a fast moving core speeding up the skin and producing narrowing instead of swelling can be discounted since the fast cooling skin increases its viscosity by several decades within fractions of a second.

Clegg considers it probable that both mechanisms play a part in *low shear rate swelling* [6]. But neither accounts for the fact that at *high shear rate* (i.e. high output rates) shorter dies show larger swelling which can only be explained as a memory effect of the disturbed flow conditions in the die entry region. In longer dies—or with longer dwell times in the same die—the memory effect diminishes. In fact the diminution of the memory effect occurs in that part of the flow curve which is below the critical point for extrudate irregularity at an approximately similar shear stress.

The comparatively mild waviness associated with flow defects below shear rates corresponding to the occurrence of gross distortion cannot be due to turbulence, since it occurs below the Reynolds number. In order to locate the seat of this behaviour Clegg studied the flow pattern by extruding rod containing a coloured core. This was achieved by filling an inverted ram extruder barrel specially constructed for the laboratory investigation with natural polythene discs whose centres had been cut out and had been replaced by a coloured polymer of the same grade.

At low output rates the extrusion of such a charge gave a uniform extrudate in which the core remained central until a fairly high output rate was reached. At higher output rates the core started to move like a vortex streamline with increasing sideways motion. *As the output rate was increased still further* towards the critical point a state was reached when no core material at all was extruded for a brief interval. A knobble or wave appeared on the extrudate when this occurred and the whole process of core–no core–core–no core was repeated regularly. This indicated that the seat of flow disturbance was in the die entry region. This was investigated further by taking ciné films through a series of glass dies fitted to the ram of the inverted ram extrusion device. Oscillatory flow of the core was observed with increasing output rate (Fig. 5.1).

Eventually the vortex-like oscillatory movement became of such an amplitude that the continuous thread of the central cone broke. The upper portion appeared to retract away from the die entry and to follow the eddies. This effect was pronounced when the die entry was square and substantially

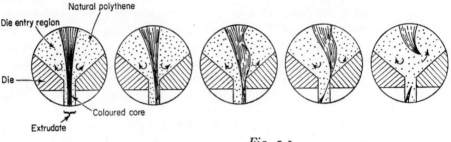

Fig. 5.1

reduced with an inlet of a 12° half-angle cone. The retraction is clearly due to the influence of the eddying currents at the inlet shoulder which were shown to be greatest with a square cut inlet by putting poppy-seed markers into the natural polythene discs. By using a 12° inlet cone half angle instead of the 90° (square cut) half angle it was possible to reduce the actual die length by over one-third and at the same time increase about tenfold the critical shear rate above which melt fracture occurred.

All this showed that undisturbed flow near the die entry is a necessary condition for regular extrudate. Irregularity only occurred when polymer came through which lay outside the die entrance streamline through-flow pattern and which was involved in the eddying currents. This would occur periodically and cause an abrupt or gradual change in the extrudate cross section. Melt fracture would therefore appear to occur not so much at the die entry proper (as Tordella believed) as some small distance upstream where the streamlines of the mainstream and of those of the eddying currents adjoin and begin to interfere with each other at higher shear rates. Relaxation times were determined from rotational viscometry. Clegg's work is neatly put in a nutshell by the table on the next page, taken from the original paper.

This indicated that polymers which show flow irregularity at low output rates are those with comparatively long relaxation times.

TABLE: Relationship between Melt Elastic and Extrusion Properties in a Polythene of Melt Flow Index 2·0 at a Shear Rate of 11·4 sec^{-1}, with a Die of $\frac{1}{16}$ in. Diameter and 0·45 in. Length at 190°C

Polythene Sample	Concentric Cylinder Rotational Viscometric Measurements at 130°C		Extrusion Measurements at 190°C	
	Relaxation time (sec)	Maximum shear stress τ (dynes cm$^{-2} \times 10^{-6}$)	Critical shear rate (sec^{-1})	Critical shear stress (dynes cm$^{-2} \times 10^{-6}$)
A	0·20	1·3	1,600 (!)	4·1
B	0·27	1·1	400	2·4
C	0·29	1·1	200	1·4
D	0·31	1·0	100 (!)	1·3

Bagley, Storey and West [7] likewise experimented with polythene and came to similar conclusions, as did Eckert [8] who worked with Neoprene.

Schreiber, Rudin and Bayley [9] have attempted to separate the elastic from the viscous effects in polymer melt extrusion, giving reasons for the different responses evoked by those effects respectively. Their rationalisation is based on flow unit flexibility and chain interaction concepts and permits of the prediction of certain experimentally verifiable aspects of polymer melt rheology.

The state of knowledge was built up as follows:

1. In calculating the shear stress at the wall of the die from

$$\tau = \frac{\Delta PR}{2L_c}$$

L_c is the effective die length obtained by making a correction for the entrance effect. This end correction is generally taken to be about 6 radii for most polymer melts, but Clegg found that it increases with increasing output rate up to the critical point and can be as large as 8 radii.

2. Bagley showed that the end correction in the rod extrusion of polythene could be divided into two components:

(a) A viscous (Couette) component.

(b) An elastic (recoverable shear) component.

Bagley also showed that the *true melt viscosity was independent of the orifice length*.

3. In a subsequent paper, Bagley et al. examined extrudate swelling in various polythenes. They found that *melt elasticity*

varied with the total shear rate and therefore *depended on the die length*.

4. Tordella concluded from birefringence measurements that the elastic strain in melt flow varied down the die channel whilst the corresponding viscosity changes were quite small. This clinched the issue: elastic parameters depend on the die length whilst viscosity does not.

5. Schreiber and Bagley found that some linear polythene fractions needed no end correction for effective die length, whilst unfractionated whole polymers of the same average molecular weight as the fractions had appreciable end corrections. The two types also differed in extrudate swelling and in the increase in swelling with applied stress: Swelling was not only much less with fractions, it also did not seem to matter whether the fractions were sharp or blended. Indeed, reconstitution of the fractions of the whole polymer still did not restore extrudate swelling. Clearly, factors other than molecular weight distribution were responsible for the differences in elasticity and these factors must arise as a direct result of the fractionation process itself. This was confirmed by Schreiber, Rudin and Bagley when they found that the viscosity of fractions was unaffected by solution treatment.

6. A pronounced effect of solution treatment on polymer elasticity was noted by Philippoff who observed a marked reduction in the birefringence of polystyrene solutions after filtration—whilst the viscosity remained unchanged. (See Chapter 1.)

This result led to the conclusion that the processes of fractionation and filtration in solution have the same effect, namely the elimination or reduction of elastic response in melts obtained from the fractions subsequently, whilst leaving the viscous response unaffected. There can only be one interpretation of the cause of this: since the size of flow units in polymer fractions and in unfractionated polymers is much the same provided the polymer is above a certain critical molecular weight, the size of the flow units will be unaffected by solution treatment. But temporary 'crosslinks' (entanglements of chain segments) which are normally present are loosened by solution treatment and 'combed out' by filtration. The reduction in the number of entanglements will cause a corresponding reduction in elasticity. One would expect that the entanglements are largely restored by keeping 'combed' polymer in the melt state over a certain period of time. It has been confirmed that *some* elasticity (manifested by die swelling) is always present, not only because the melt exists over a fair time interval before it reaches the die, but also because the flow disturbance at the die entrance promotes fresh entanglement. As the melt flows down the die itself the flow will return to regularity, some disentanglement may occur and the elastic effects will settle down to an unspectacular steady state condition.

Clearly, for network entanglement to occur at all and for elasticity to manifest itself, it is necessary that the polymer chains should be of a certain critical length. This critical length has been calculated by S. J. Gill and R. Toggenburger [9a] from birefringence data. They showed it to be at least twice the critical length at which the maximum flow unit length has been reached.

Steady states in long range entanglements are attained slowly in polymer melts under the usual conditions. Melt elasticity therefore depends greatly on the previous history of the polymer including:

(i) Solution treatment and filtration—which will loosen and 'comb out' entanglements;

(ii) turbulent shear which will exaggerate entanglements but which may also cause an opposite effect if applied in excess because of mechanical fragmentation of chains.

(iii) thermal history which, like excessive mechanical working, will cause degradative shortening of chains.

The greater the molecular weight the more important will the polymer history be on the elastic extrudate properties. But the viscous properties should be nearly independent of the polymer history since the entanglements are not likely to affect the flow unit size, although they may have an influence on the free volume of the melt.

Schreiber and Storey [10] have taken up a suggestion of Busse [11] that molecular fractionation could occur along the wall of the die and consider that molecular fractionation during flow through the die contributes to the reduction in extrudate swelling.

The probable slip-stick effect of melt at the wall of the die and its significance for minor melt irregularities as suggested by Tordella have already been mentioned. Direct evidence for the existence of discontinuities caused by the wall has come from a paper by Galt and Maxwell [12] who, like Clegg, used a particle tracer technique but concerned themselves more particularly with flow patterns along a transparent die channel wall, taking photographs through a camera which was focused on various points along the radius of the circular channel. The work was done on low density polythene melts ($d = 0.92$ to 0.933) with melt flow indices from 0.19 to 19.5, branching of from 6 to 37 methyl groups per 1,000 carbon atoms and weight average molecular weights from 140,000 to 500,000.

Galt and Maxwell found that the melt velocity at the wall was not *invariably* zero and that slip-stick effects were therefore inevitable. The zone where this occurred could represent as much as about one-sixth of the channel radius. The boundary between the wall zone and the remainder of the channel section showed an abrupt transition in the flow pattern which became more pronounced and extended deeper towards the axis as the shear stress was increased. This boundary was additional to another boundary still further towards the axis which was not abrupt but which marked the extent of a central core moving as a plug with constant velocity. They attribute the wall boundary annulus region 'to the highly elastic behaviour of polymer melts that do not flow continuously. The highly elastic melt elements are created by long-chain branching which causes the generation of molecular clusters.'

In other words, they detected an element of dilatancy in what is considered to be a 'typically pseudoplastic' flow system and they detected it in the region of the die where the shear stress is highest. This links up with the point of view put forward in Chapter 1 that a generalised flow curve will have a dilatant portion at high shear stresses. The fact that this could be experimentally observed must be ascribed to the experimental technique which constrained the pressurised flow by means of the die geometry and so prevented turbulence. Whether 'structure' can be created in the same way with *perfectly* linear polythene (polymethylene) would therefore be a matter of great fundamental interest. If the same effect were observed to a lesser degree, then this would mean that chain branching is conducive to the formation of shear 'structure' at high shear stresses—which has been postulated in Chapter 1. Some evidence in this respect is already available. Film extruded from high density polythene has less *surface* haze due to small-scale extrusion defects than has low density polythene film extruded under corresponding conditions of melt viscosity and shear rate.

On the other hand, Galt and Maxwell's 'molecular clusters' may be identical with the 'crystallites' which Van der Vegt and Smit have reported in the melt stage of certain polymers under fairly high shear rate conditions in capillary channels and to which attention has already been drawn on several occasions. It is therefore possible to suggest that melt fracture may be due to melt crystallites forming at the die entrance in a discontinuous fashion, depending on local variations and oscillations in shear rates throughout this zone. The fact that melt fracture occurs in both 'crystalline' and 'amorphous' polymers does not invalidate but, on the contrary, supports this argument, since 'melt crystallinity' has been reported even in such typically amorphous polymers as polystyrene and PVC.

The question as to whether the melt instability which ultimately leads to melt fracture arises in the die entry region or as a result of slip-stick effects in the die itself has already been discussed. As a consequence of work published only days before the manuscript for this volume reached the proof stage it is now recognised that both play a part and that the effect, when it is observed, is a delayed reaction to the first.

F. N. Cogswell and P. Lamb [15] investigated the criterion for the 'triggering' of slippage, starting from a consideration of the end correction for dies which may be written:

$$\tau_{\text{true}} = (P - P_0)r/2l,$$

where P_0 is a function of the true shear stress and is largely independent of temperature. A graph of P_0 vs shear rate showed a discontinuity in an otherwise linear plot at some critical shear rate at which distortion commences.

If this distortion is due to slip-stick effects, as J. J. Benbow and P. Lamb suggested [13], then the balance of axial forces in the die wall region may be considered with profit. The elastic stress which is a function of P_0—say $f_1(P_0)$—is always greater than the shear stress at the wall [14]. There will also be a static adhesion term $f_2[(P - P_0)l'/l]$ which will be a function of the pressure at the particular distance from the die exit and which will be a minimum when l' is zero—i.e. at the die exit itself. $f_1(P_0)$ may also be a function of l' due to the relaxation of elastic strain during the passage of melt through the die. There may be additional forces Z acting from outside the die in the flow direction (e.g. haul-off). Cogswell and Lamb suggest that slip-stick is triggered off when

$$f_1(P_0) - \tau > f_2[(P - P_0)l'/l] + Z,$$

and they go on to amplify this criterion as follows:

'The initial triggering will occur preferentially when the adhesion term is a minimum, i.e. at the die exit where l' is zero. However, P_0 increases more rapidly than τ, and at higher shear rates will be able to overcome greater adhesion; because of this, the site of triggering will progress back towards the die entry as throughput is increased beyond the critical value. When this slipping occurs P_0 relaxes slightly, allowing the polymer to re-adhere to the die wall; this results in a periodic effect. The periodicity and mass per cycle M are related to the volume over which elastic strain is relaxed as a result of one initiating slip event.'

Cogswell and Lamb use the mass per cycle M as a measure of the severity of distortion and suggest that irregular distortion occurs when the trigger site has moved up to the die entry $(l' \equiv l)$, in which case, by the above-mentioned criterion $f_1(P_0) - \tau$ must inevitably be greater than $f_2[(P - P_0)l'/l] + Z$, unless Z is unusually large (e.g. the haul-off in spinning dies). For a constant $\dot{\gamma}$ and constant l/r, M is proportional to r^3 and this implies that, for a fixed value of P_0 the volume of melt relaxed by a single slip-stick cycle occupies the same proportion of the die volume, whatever the die diameter. In other words M/r^3 constitutes a normalised measure of severity, so that the distorted extrudate from a small die is a true mini-reproduction of the distorted extrudate from a large die. M was plotted vs τ for polypropylene and showed a smooth curvilinear relationship which was, moreover, little affected by degradative treatment of the polymer, that is to say, by a reduction in molecular weight. This meant that volume and stress distribution effects are more important than molecular relaxation phenomena.

Another additional effect to which Cogswell and Lamb have drawn attention is that the mass per cycle decreases as the piston approaches the die:

'At small piston heights there is an increase in *radial* flow and the layer which must pass through the die closest to the die wall receives a greater acceleration than under normal conditions and will therefore be more likely to trigger distortion. While the localised boundary layer receives greater acceleration (elastic strain) the central portion receives less, and there is no change in the total pressure drop or output rate. This effect was investigated as a function of piston height and thermal history which in no way affects the result. The evidence again points to the die entry as the region where the bulk effects responsible for distortion are introduced; triggering takes place in the boundary layer at the polymer/metal interface where maximum elastic strain occurs.'

The conclusions which are finally reached may be summarised as follows:

(1) The onset and severity of distortion depend on the recovery of elastic deformation;

(2) The elastic deformation is imparted to the polymer melt during acceleration at the die entry;

(3) Recovery is triggered by slippage at the die wall in a direction opposite to the direction of flow;

(4) The recovery may be triggered where adhesion is lowest, i.e. at or near the die exit;

(5) The recovery may conceivably be prevented and melt distortion avoided by exerting large Z forces (high differential drawdown).

A most important corollary of this work arises which will be more fully appreciated after reading the chapters dealing with deformation in the solid state. It is, however, necessary to develop some of the arguments at this point:

Since melt instability involves a mass-accelerational term we are in fact dealing with a situation which is fundamentally dynamic and which must therefore lead to sinusoidal oscillation. The period (reciprocal frequency) of this oscillation is directly related to the applied shear stress. It is also inversely related to the die length: The longer the die, the longer will be the dwell time of the melt in the die and the greater will be the probability that the dwell time in the die will fall within the limiting time required for relaxation to occur on a gradual (molecular) microscopic scale rather than in macroscopic (slip-stick) steps. The application of Z-forces (hot drawing) has the nett effect of increasing the effective die length because this imposes restrictive boundary conditions beyond the die exit.

If it is accepted that we are facing a fundamentally dynamic situation under all extrusion conditions, then we can distinguish three special cases which differ amongst themselves only in the relationship between the mass-accelerational term on the one hand and the combined melt and die characteristics on the other:

(1) The mass-accelerational term is negligibly small and any sinusoidal oscillations induced at the die entry are damped out during the passage of the polymer through the die because the channel is sufficiently long to necessitate a throughput time well in excess of the relaxation time of the viscoelastic melt. As a consequence the oscillatory period becomes zero and the extrudate is smooth.

(2) The mass-accelerational term is so large that the sinusoidal oscillations are no longer completely damped out. Depending on the magnitude of the term, the period of oscillation will increase and with it the mass per cycle M, leading to surface mattness, sharkskin, bambooing and porpoising in the extrudate.

(3) The mass-accelerational term has become so large that the continuity of the process, that is to say, its periodicity, is lost altogether; melt fracture is occurring all the time. This situation is directly comparable to that which would arise when a normally sinusoidally vibrating

solid has been given an initial deformation which stresses that solid beyond its elastic limit; fracture or permanent deformation occurs and any observations on the subsequent dynamic responses become meaningless. The emerging polymer loses its periodicity and appears as a useless stream of extrudate the profile of which is irregular and shows little (if any) resemblance to the geometry of the die exit.

It is therefore seen that a 'static' situation (smooth extrudate) is merely a special case of a generalised dynamic situation in which the mass-accelerational term may be neglected owing to the fact that in comparison to the viscous and elastic deformation it becomes vanishingly small. Whilst this applies to the liquid state here, a precise analogue exists in the solid state, as will be seen in the following chapters. It will be clear, therefore, that we have here another facet of the fundamental analogy between the so-called 'solid' and 'liquid' states.

6 Rheology applied to extruder die and screw design

Rectangular dies, wide-slit dies and annular dies as the only dies suitable for analysis by currently available and virtually rigorous mathematical methods. Examples of suitable dies based on power-law materials. Annular dies treated as wide-slit dies. Worked examples (after Carley). Extruder equations and screw-die interactions.

IN considering flow problems it has so far been assumed that we are dealing with a simple circular die where the velocity is the same for any circle concentric with the axis of the extrusion cylinder. The circles resemble the contour lines on a map and join points of the same velocity [1]. These lines are termed 'isovels'. If the channel is, however, rectangular, then the isovels tend to become elliptical towards the centre (Fig. 6.1).

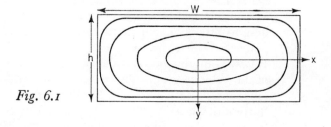

Fig. 6.1

The velocity cannot be defined at any one point in this cross section with a single coordinate. Some symmetry obviously does exist and the velocity distribution in any one quadrant is mirrored in any other. But in that quadrant the velocity varies not only with the distance from the centre but also with the angle above the horizontal axis *X*. Since the boundaries of the channel are parallel to the *X*, *Y* axes it is convenient to give the quadrant *XY* coordinates and to specify the velocity at any point in terms of these coordinates. The flow is therefore *two-dimensional* whilst in a round hole it is *unidimensional*.

If the rectangular cross section, however, is much wider than it is high ($w \gg h$), then the isovels are parallel to the slit direction over most of the slit width and the end regions (where they are not) constitute only a small part of the entire cross section (Fig. 6.2).

In die design work it is very important to recognise in what kinds of cross sections two-dimensional flow can be approximated by unidimensional flow.

Calculations of flow rates for power-law liquids in terms of flow rates can be made for all unidimensional flow geometries.

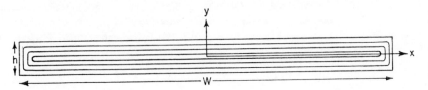

Fig. 6.2

It was seen in Chapter 3 how the relevant equations are derived and manipulated. But when it comes to two-dimensional geometries, then only a few of the geometries have equations derived for them (rectangular, elliptical and equilaterally triangular) and even then these can only be used for Newtonian liquids. Solutions of the highly complex differential equations for other geometries and for power-law liquids in general have not yet been found.

Another factor which complicates the calculations for die design for polymer melt extrusion is the melt elasticity of the material which, as we know, is responsible for die swelling. If flow is unidimensional the swelling is mostly in the direction of the velocity gradient; a circular die will still give a circular extrudate, and a flat strip leaving a slit die will swell mostly in the thickness direction and only slightly in width. But when flow is two-dimensional as, for instance in a square or rectangular section die the shear stress will be greatest parallel to the sides and least along the diagonals, so that swelling will be great at the side centres and small at the corners. The shape of the extrudate will therefore tend towards a circle in a square die and to an ellipse in a rectangular die. For the present therefore we have to content ourselves with the quantitative treatment of general power-law liquids in dies which have *unidimensional geometries*. Nevertheless, as will be seen presently, much can be achieved even within this limitation.

Many dies are designed today without any quantitative consideration of the flow of the melt in the channel. An attempt is usually made to 'streamline' the flow passages in order to avoid noticeable polymer degradation, but efforts to design the die so as to deliver a given output of the proper profile with some specified applied pressure are not often made. Instead people tend to adopt an empirical attitude which can

·waste both time and material. The trade extruder who can apply quantitative flow principles to die design will therefore enjoy the advantage of greater competitiveness.

Flow curves

All flow calculations related to die design must be based on the flow behaviour of the resin. The best form of these for the die designer is that of a log/log plot of apparent shear rate vs shear stress at several temperatures within the processing range. A good collection of plastics flow data is found in the charts of Section III of *Processing of Thermoplastic Materials* (Bernhardt, Editor, Reinhold N.Y. 1959). The curves for other thermoplastics are broadly similar. Their slope may vary, they may be spaced closer or wider, etc., but they do have certain common features:

    1. They slope a little more steeply at higher shear stresses and the slope does not change much from one temperature to another.

    2. They are concave upward.

    3. They can be approximated (over a reasonable range at least) by straight lines.

    4. Shear rates always increase with rising temperature at a given shear stress.

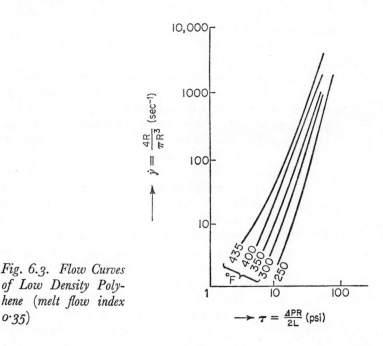

Fig. 6.3. Flow Curves of Low Density Poly-hene (melt flow index 0·35)

These similarities form the basis for a single quantitative method for die design that applies to all commerical polymers.

The power-law equation:

Equation 6.1
$$\dot{\gamma} = \frac{4Q}{\pi R^3} = k'\left(\frac{\Delta PR}{2L}\right)^m = k'\tau^m$$

where $m = 1/n$ is equivalent to the statement that the log/log flow curves are replaceable with straight lines.

The exponent m is the tangent slope of the line, k' is the shear rate at a shear stress of 1 psi (intercept). The slope is generally between 1·5 and 4. To calculate m and k' let us take the 250°F line approximating the experimental curve in Fig. 6.3. Taking two points on that line from the graph:

$$\begin{cases} \tau_1 = 10, & \dot{\gamma}_1 = 1\cdot25 \\ \tau_2 = 100, & \dot{\gamma}_2 = 2{,}000 \end{cases}$$

and substituting in the power-law equation we get:

$$\left.\begin{array}{l} 2{,}000 = k'(100)^m \\ 1\cdot25 = k'(10)^m \end{array}\right\}$$

$$\therefore \quad \frac{2{,}000}{1\cdot25} = \frac{100^m}{10^m} \quad \text{or} \quad 1{,}600 = \left(\frac{100}{10}\right)^m = 10^m$$

whence $m = 3\cdot20$, and substituting in either of the above pairs (say: $2{,}000 = k'(100)^{3\cdot20}$, therefore $k' = 8\cdot0 \times 10^{-4}$).

Even if the flow curve bends rather more than one would like for fitting a line, one can take a line in the high shear rate region since the shear rates in normal dies (except for certain pipe and cable coverings) are normally high. Whatever the appropriate range of shear rate—the line is fitted to that range.

True unidimensional flow exists only in circular channel dies and in wide thin slits without ends. An interesting and practically important special case of both of these types of unidimensional flow are dies whose openings are circular annuli (for pipe, wire coating and blown film). On the other hand, the annulus can also be regarded as an endless slit obtained by bending the slit round a centre. It differs from the infinite slit in only one respect: The ID < OD, so that the effective slit width is somewhere between $2R_o\pi$ and $2R_i\pi$, R_i and R_o being the outer and inner radius respectively.

The flow in an annulus has been analysed exactly for Newtonian and certain power law liquids where m is a whole number. The flow rates calculated for given annulus dimensions and pressure drops from the exact analysis agree very closely with those from a simple approximate formula obtained by treating the annulus as a wide slit, except where $OD \geqslant 3$ ID, i.e. in exceptionally heavy walled pipe. Furthermore the approximate formula is also easy to use when m is *not* a whole number (which is usually the case). The shear rate and shear stress ($\dot{\gamma}$ and τ) given in the power-law equation are actually approximate values pertaining only to the fully circular orifice geometry and to one particular radius in the stream (the outside radius).

In actual fact, when a melt flows through a channel the shear rate and shear stress ($\dot{\gamma}$ and τ) vary from zero at the centre point to its maximum at the die wall. The value at the die wall ($\dot{\gamma} = 4Q/\pi R^3$) is true only for Newtonian liquids.

For power-law liquids the 4 must be replaced by $(n + 3)$ to give the true shear rate at the wall, which is $\dot{\gamma} = (n + 3)Q/\pi R^3$. However, the 'apparent' shear rate $4Q/\pi R^3$ is very handy to work with since it can be computed before m is actually known. τ is $\Delta PR/2L$, irrespective of the material.

Both shear rate and shear stress at the wall are different when switching from the circular to the slit geometry. The stress at the wall of a wide thin slit is $\Delta Ph/2L$, where h is the slit thickness.

The shear rate for Newtonian liquids (i.e. the *apparent* shear rate for non-Newtonian liquids) at the slit wall is $\dot{\gamma} = 6Q/wh^2$, where w is the width of the slit. These differences in geometry, and the fact that the apparent shear rate is a geometry-dependent approximation, makes the power-law equation a little different for slits from that applicable to circular dies:

Equation 6.2
$$\dot{\gamma} = \frac{6Q}{wh^2} = k'' \left(\frac{\Delta Ph}{2L} \right)^m = k'' \tau^m$$

k'' is slightly smaller than k' and the ratio of k''/k' varies slightly with the value of m. The variation, however, is small enough to take $k''/k' = 0.91$ for all values of m with little error.

The two power-law equations 6.1 and 6.2 (for circular and slit dies respectively) can be rewritten:

Equation 6.3
$$Q = \frac{\pi k' R^{m+3} \Delta P^m}{2^{m+2} L^m}$$

and

Equation 6.4
$$Q = \frac{k'' wh^{m+2} \Delta P^m}{3(2^{m+1}) L^m} = \frac{k' wh^{m+2} \Delta P^m}{3 \cdot 3(2^{m+1}) L^m}$$

The flow rate unit of Equations 6.3 and 6.4 in volume rate of flow is not very convenient for die design work. Extruders think in terms of mass rate of flow. To convert, one must know the melt density under the conditions in the die. Such information is scarce, but Bernhardt (loc. cit.) is again the best source. In the absence of such data, the following approximation can be used:

Equation 6.5
$$Q_m \cong 115\rho_0 Q$$

where Q_m = mass rate of flow in lb/hr,

ρ_0 = specific gravity at room temperature,

Q = volume rate of flow, cu in./sec.

Where the actual melt density ρ_m in g/cc is known, the proper equation to use is

Equation 6.6
$$Q_m = 130\rho_m Q$$

Equation 6.5 can now be combined with Equations 6.3 and 6.4 to give the working output equation for practical die design:

For circular channel dies:

Equation 6.7
$$Q_m = \rho_0 \left(\frac{90k' R^{m+3} \Delta P^m}{2^m L^m} \right)$$

For wide-slit dies:

Equation 6.8
$$Q_m = \rho_0 \left(\frac{17 \cdot 4k' w h^{m+2} \Delta P^m}{2^m L^m} \right)$$

When is a wide slit wide enough?

The error is:

$$\sim 10 \text{ per cent} \quad \text{when} \quad w/h = 7$$
$$\sim 7 \text{ per cent} \quad \text{when} \quad w/h = 10$$
$$\sim 5 \text{ per cent} \quad \text{when} \quad w/h = 13$$

Circular annuli can also be treated as wide thin slits. Equation 6.8 can be used, but for the annulus

Equation 6.9
$$w = \pi(R_o + R_i)$$

and

Equation 6.10
$$h = (R_o - R_i)$$

where R_o and R_i are the outer and inner radii respectively. Thus, the effective width of the annulus is just the circumference at the mid point.

Many profile dies can be thought of as a combination of wide thin slits and/or annuli. The annuli need not be circular. If the slit thickness is the same throughout, the only problem is to determine the proper width w to use in Equation 6.8. Examples of typical shape dies in which flow rates are calculable from slit-die formulae include the following (Fig. 6.4):

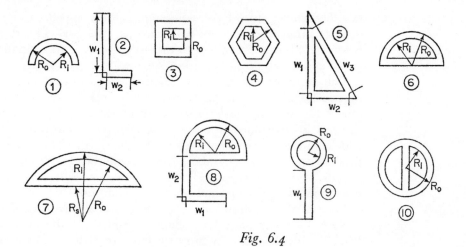

Fig. 6.4

Shape 1 is a half-annulus. Unlike the complete annulus, this *does* have ends, therefore w/h should be $\geqslant 10$ if Equation 6.8 is to be used without end-effect shape factor. The effective width is one-half the full annular width, or $\pi/2(R_o - R_i)$.

Shape 2 also has closed ends, but only one end is closed for each branch.

Shape 3 is a square annulus without ends. The appropriate w is $4(R_o + R_i)$.

The appropriate w values for Equation 6.8 are:

Shape	
1	$\pi/2(R_o - R_i)$
2	$w_1 + w_2$
3	$4(R_o + R_i)$
4	$3\cdot46(R_o + R_i)$
5	$w_1 + w_2 + w_3$
6	$\pi/2(R_o + 3\cdot57R_i)$
7	$(R_o + R_i)(\theta + \sin\theta)$
	where θ (in radians) is the angle whose cosine is $2R_s/(R_o + R_i)$
8	$\pi/2(R_o + 3\cdot57R_1 + w_1 + w_2)$
9	$\pi(R_o + R_i) + w_1$
10	$\pi R_o + 5\cdot14R_i$

Entrance corrections

The flow formulae (Equations 6.7 and 6.8) imply a direct proportionality between pressure drop and land length in any die. It has always been supposed that shorter lands require lower pressures. But rheologists have found that in successive tests on orifices with decreasing lengths, there appears to be a definite pressure drop even with zero land length. This drop is *not small*, but is equivalent to the drop observed in a circular die of actual land from 1 to 6 diameters.

This entrance loss is usually explained in terms of the energy needed to change the stream suddenly from a fat sluggish one to a thin fast one with stretched molecules. But some of this loss is occasioned in forcing the melt to flow through the wide chamber upstream of the actual orifice itself. The pressure drop in the supply chamber, while negligible in Newtonian liquids, is by no means negligible in polymer melts which are pseudoplastic under normal processing conditions.

Most of the published data giving $\dot{\gamma}$ and τ have not been corrected for entrance loss. Bagley found that the effective length equivalent to the entrance loss varied with conditions but average between two and three orifice diameters. Ryder, after examining a number of polymers, obtained effective entrance lengths ranging from one to five diameters. In the absence of specific information, let us therefore assume the entrance length to be three diameters (six radii).

This assumption leads to simple formulae for correcting shear stresses and die lengths.

$$\tau \text{ (for circular orifices)} = \frac{\Delta PR}{2L}$$

To correct this for entrance loss one must multiply by

$$\frac{L/R}{L/R + 6}$$

This is also useful in adjusting calculations for the difference between the L/R ratio of the rheometer used by the rheologist in the flow curve determination and the L/R ratio of the projected die.

Example

Data and Problem

It is desired to extrude a $\frac{1}{4}$ in. round rod from the polythene whose flow curves were given earlier on.

To facilitate cooling and shape holding it will be wise to use a relatively low temperature, 250°F at which the melt is to pass through the die. The desired production rate is 52 lb/hr (23·6 kg/hr) and the rod is to be drawn away from the die at just the average velocity with which it flows through that die—i.e. with just enough pull to compensate for the difference between viscoelastic die swelling and the contraction on cooling. The nett draw ratio

$$\left(\text{the ratio of }\frac{\text{die cross section}}{\text{cross section of finished extrudate}}\right)$$

is 1·00. The density of the melt (from Bernhardt) is 49·6 lb/cu ft (0·795 gm/cc) = 0·0287 lb/cu in. The flow rate in cu in./sec is 52/3,600 × 0·0287 = 0·504 (8·25 cc/sec).

Past experience with the 2-in. extruder to be used for the job indicates that it will deliver 52 lbs/hr of low density polythene of melt flow index 0·35 most reliably against a head pressure of ∼1,200 psi (84 kg/cm²).

What should the land length of the die be?

Solution

We know the melt density in this case and can therefore use Equation 6.3:

$$Q = \frac{\pi k' R^{m+3} \Delta P^m}{2^{m+2} L m}$$

k' and m are known from the available flow curve at 200°F: $k' = 7·9 \times 10^{-4}$; $m = 3·20$.

Solving Equation 6.3 for L we get:

Equation 6.11
$$L = \left[\frac{\pi k' R^{m+3} \Delta P^m}{2^{m+2} Q}\right]^{1/m} = \left[\frac{\pi k'}{Q}\right]^{1/m} \frac{R^{(m+3)/m} \Delta P}{2^{(m+2)/m}}$$

R is given as $\frac{1}{8}$ in., ΔP as 1,200 psi, Q is 0·504 cu in./sec.

$$L = \left[\frac{3·14 \times 0·00079}{0·504}\right]^{1/3·20} \times \frac{0·125^{6·20/3·20} \times 1,200}{2^{5·20/3·20}} = 1·32 \text{ in.}$$

$$L/R = 10·5$$

So far no effort has been made to correct the pressure drop for entrance losses.

The original flow measurements were made on a die of $L/R = 18·3$. The present L/R is considerably less, so corrections should be made. The *effective* pressure drop in the rheometer orifice would have been 18·3/18·3 + 6 = 0·753 of the actual pressure drop.

For our short die, however, the effective pressure drop would be only 10·5/10·5 + 6 = 0·666 times the actual pressure drop.

Since the pressure drop is in direct proportion to land length, the proper correction to apply to the land length calculated above is just the ratio of these two correction factors, or

Equation 6.12

$$L_{corrected} = L_{nominal} \times \frac{b(a + 6)}{a(b + 6)}$$

where $a = L/R$ for the die of the rheometer used in obtaining the flow curve $b = L/R$ for the uncorrected die.

For our example, $a = 18\cdot3$, $b = 10\cdot5$,

$$L_{corrected} = 1\cdot32 \frac{10\cdot5 \times 24\cdot3}{18\cdot3 \times 16\cdot5} = 1\cdot12 \text{ in.}$$

By neglecting the entrance correction, the land is made about 18 per cent too long in this case. This would mean that only 31 lbs/hr would be obtained with a head pressure of 1,200 psi, or that, conversely, the head pressure would have to be raised to 1,420 psi to realise the desired 52 lbs/hr output. This could be done by increasing the screw speed, provided that this change does not cause other difficulties (feeding troubles, overheating).

The shorter the die (the lower L/R), the more important is it to apply corrections.

Similar corrections can be applied to slit dies by treating the L/h ratio in the same way as the L/R ratio was treated for round dies.

A number of excellent worked examples has been given by Carley [2] and the reader is strongly recommended to examine these closely. In yet another paper Carley deals with the design and operation of sheeting dies [3], gives a mathematical analysis of sheeting and blown film dies and climaxes this by again providing an illustrative worked example.

It would be possible to write a sizeable book on the subject in which Carley and others are active and perhaps this may yet be done. The mathematics involved are rather complex. An abbreviated treatment of extruder equations and screw-die interactions is given below and the reader is advised to consult the following references, should he require a more detailed mathematical analysis:

1. McKelvey [4], *Polymer Processing*, Ch. III, Sections 5 to 8 including, especially, wire coating dies, pp. 111–113;

2. Mori and Matsumoto [5], *Rheol. Acta*, 1958, **1**, 240.

3. Griffiths [6], *Ind. Eng. Chem.* (Funds.), 1962, **1**, 180.

Calendering has been considered on the basis of similar mathematics. The original article analysing calender flow appeared in 1950: R. E. Gaskell [7], *J. Appl. Mech.*, 1950, **17**, 334.

The treatment assumed Newtonian flow but can be extended to generalised power-law liquids, so long as it is confined to a unidirectional flow pattern. This was done and is covered by

McKelvey (*Polymer Processing*) in Chapter IX of that splendid book.

Extruder equations and screw-die interactions*

The following treatment of extrusion is much condensed and therefore appropriate to the scope of the present volume.

Assume that we are dealing with a screw having a relative velocity with respect to the barrel of N r.p.m., a constant channel with w which is very much greater than the constant channel depth h, a flight angle α and flight thickness t and a negligibly small flight clearance. This screw is moving in a barrel of diameter D which is very much greater than the channel depth h. The coordinates are chosen so that x is the direction of the channel width, y is the direction of the channel depth and z is the axial direction of the channel or the main flow direction. The velocity components in these directions are v_x, v_y and v_z respectively.

In a wide slit the end effects of the channel (at the flights) can be ignored because the isovels are essentially parallel. It does not matter whether we consider the screw rotating in the barrel or the barrel rotating round a stationary screw since the relative motion alone controls the output of the screw pump. For easier manipulation it is convenient to formulate our equations on the basis of a stationary screw. We can reasonably assume that flow in the y direction is negligibly small. The flow in the x direction cannot, strictly speaking, be ignored. It varies with the positional coordinate y from a maximum value at the moving boundary $(y = h)$ given by

$$v_x = \pi D N \sin \alpha$$

to a minimum value at the base of the channel $(y = 0)$ which is given by

$$v_x = 0$$

The velocity in the z direction is approximated by the parallel plate model (see Chapter 3) and is, again, only a function of the positional coordinate y/h, so that both v_x and v_z are a function of the channel depth only:

$$v_x = f(y)$$
$$v_y = f(y)$$

* I am indebted to Dr. Zamodits whose Ph.D. thesis at Cambridge used the approach which follows.

The rheological equation $\tau = \eta\dot{\gamma}^n$ can be written:

Equation 6.13
$$\tau = \eta\dot{\gamma}^n = c(\dot{\gamma}^2)^{-s}\dot{\gamma}$$

where c is a constant and $s = (1 - n/2)$.

The shear rates for flow in the x direction and z direction are deformations acting perpendicularly to each other and are given respectively by:

$$\dot{\gamma}_x = \frac{dv_x}{dy} \quad \text{(let this be } U_y)$$

and

$$\dot{\gamma}_z = \frac{dv_z}{dy} \quad \text{(let this be } W_y)$$

The viscosity of the liquid is, however, dependent on the *total* shear rate $\dot{\gamma}$, so we must add the individual directional shear rates vectorially, or

$$\dot{\gamma}^2 = U_y^2 + W_y^2$$

By definition $\eta = \tau/\dot{\gamma}$ and substituting for τ in the rheological equation we get

$$\eta = c(\dot{\gamma}^2)^{-s}$$

It is now necessary to resolve the stress tensor $\boldsymbol{\tau}$ into its components. We can neglect the component in the y direction on the previously made basic assumption that flow in the y direction is negligibly small.

The components in the x direction are not necessarily negligible and are included in this treatment. Their derivatives with respect to the axis along which the force is applied (x in τ_{xx}, y in τ_{xy}, z in τ_{xz}) must sum to zero because nothing in the profile except the pressure depends on the distance along the channel z. Hence:

$$\frac{d\boldsymbol{\tau}}{dz} = 0 = \frac{\partial \tau_{xx}}{\partial x} + \frac{\partial \tau_{xy}}{\partial y} + \frac{\partial \tau_{xz}}{\partial z}$$

Now τ_{xx} is equal to the normal force applied to the system, i.e. the pressure, or

$$\tau_{xx} = -P$$
$$\tau_{xy} = \eta\dot{\gamma}_{xy} = c(U_y^2 + W_y^2)^{-s}U_y$$
$$\tau_{xz} = \text{negligibly small} \cong 0$$

Substituting for the components in the previous equation we therefore obtain:

$$\frac{\partial(-P)}{\partial x} + \frac{\partial[C(U_y^2 + W_y^2)^{-s}U_y)}{\partial y} = 0$$

or

Equation 6.14
$$\frac{\partial P}{\partial x} = \frac{d}{d_y}[C(U_y{}^2 + W_y{}^2)^{-s}U_y]$$

By exactly the same reasoning we obtain the analogous equation for the components of the stress tensor τ in the z direction which results in the equation:

Equation 6.15
$$\frac{\partial P}{\partial z} = \frac{d}{d_y}[C(U_y{}^2 + W_y{}^2)^{-s}W_y]$$

We have already stated that the boundary conditions for v_x are

$$v_x(y = 0) = 0$$
$$v_x(y = h) = \pi DN \sin \alpha$$

For v_z the boundary conditions are analogously:

$$v_z(y = 0) = 0$$
$$v_z(y = h) = \pi DN \cos \alpha$$

Now, we may know $\partial P/\partial z$ from experiment, but we have no means of measuring $\partial P/\partial x$. So, in order to make use of Equations 6.14 and 6.15 we need another (additional) boundary condition. This is supplied by the fact that the existence of flights prevents the escape of liquid (nett loss) in the x-direction. We can express this by the equation:

Equation 6.16
$$\text{Output in the } x\text{-direction } (q_x) = \int_0^h v_x \, dy = 0$$

On integrating Equation 6.14 we get:

Equation 6.17
$$\int \frac{\frac{\partial P}{\partial x}}{C} \, dy = (U_y{}^2 + W_y{}^2)^{-s}U_y = \frac{\frac{\partial P}{\partial x}}{C}(y - y_1)$$

where y_1 is that level in the channel where $\partial v_x/\partial y = 0$.
On integrating Equation 6.15 we get, analogously:

Equation 6.18
$$\int \frac{\frac{\partial P}{\partial z}}{C} \, dy = (U_y{}^2 + W_y{}^2)^{-s}W_y = \frac{\frac{\partial P}{\partial z}}{C}(y - y_2)$$

where y_2 is that level in the channel where $\partial v_x/\partial z = 0$.
Dividing Equation 6.17 by Equation 6.18 we get:

Equation 6.19
$$\frac{U_y}{W_y} = \frac{\frac{\partial P}{\partial x}}{\frac{\partial P}{\partial z}}\frac{y - y_1}{y - y_2}$$

where

$$\frac{\partial P/\partial x}{\partial P/\partial z}$$

is a dimensionless quantity, denoted in the following by P_1.

Further dimensionless quantities may be formulated as follows:

$$Y = \frac{y}{h} \qquad v_x{}^* = \frac{v_x}{\pi\, DN \sin \alpha}$$

$$Y_1 = \frac{y_1}{h} \qquad v_z{}^* = \frac{v_z}{\pi\, DN \cos \alpha}$$

$$Y_2 = \frac{y_2}{h}$$

$$G = \frac{\partial P/\partial z}{C(\pi\, DN)^{1-2s}}\, h^{2-2s}$$

or, after resubstituting $1 - n/2$ for s:

$$G = \frac{(\partial P/\partial z)h^{1+n}}{C(\pi\, DN)^n}$$

If these dimensionless quantities are introduced into Equations 6.17 and 6.18 one eventually obtains after some laborious manipulation:

Equation 6.20
$$v_x{}^* = \frac{1}{\sin \alpha}\, G^{1/1-2s} \int_0^y P_1(Y - Y_1)F(Y)\, \mathrm{d}Y$$

and

Equation 6.21
$$v_z{}^* = \frac{1}{\cos \alpha}\, G^{1/1-2s} \int_0^y P_1(Y - Y_2)F(Y)\, \mathrm{d}Y$$

where

$$F(Y) = [P_1{}^2(Y - Y_1)^2 + (Y - Y_2)^2]^{s/1-2s}$$

Substituting the boundary conditions into Equations 6.19, 6.20 and 6.21, we then have three equations in terms of P_1, Y_1 and Y_2 which can be readily solved by computer.

$v_x{}^*$ and $v_z{}^*$ follow and hence v_x and v_z.

The total output of the screw is given by

$$Q = \frac{q}{Wh\,\pi\, DN}, \text{ where } q = W \int_0^h v_z\, \mathrm{d}y$$

The constants C and s are obtained from rheological data and the determination of the pressure drop $\partial P/\partial z$ defines G.

Numerical analysis by computer then enables one to plot the 'screw characteristic' (output Q vs pressure drop, or, more conveniently, output Q vs the defined value of G which is a function of the pressure drop and the power index n). All the other magnitudes involved (D, N, α, w, h) are design constants and the resulting family of curves, with each curve representing a different value of n, is shown in Fig. 6.5 which applies to a certain grade of polythene and a screw of flight angle $\alpha = 30°$:

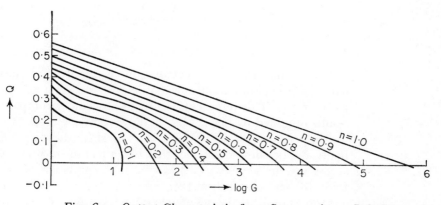

Fig. 6.5. Output Characteristic for a Screw, using a Polythene and a Screw of Flight Angle $\alpha = 30°$.

It is obvious that the assumption of $n = 1$ will give highly erroneous results, in view of the fact that n quite commonly has values between 0·3 and 0·7. The pressures necessary for polymer melts to give a certain screw output is in fact much less than that which would be needed if the melt were Newtonian, because of the pseudoplastic nature of polymer melts in general.

For the die, exactly the same procedure can be used. It is much simpler because fewer variables are involved. Again, a family of die characteristics can be computed in terms of Q vs G and although all curves will pass through the origin the curves will all deviate increasingly from linearity as n decreases from 1·0.

The point of intersection of screw and die characteristics A will indicate the operating output of the extruder (Fig. 6.6). If a different (smaller) die is substituted it is often found that the output as indicated by point B is but little affected. This is due to the fact that the screw characteristic is rather flat at that particular value for G, so that ΔQ is small. Only in a

Newtonian liquid would the change in die cross sectional area cause a strictly proportional change in nett output.

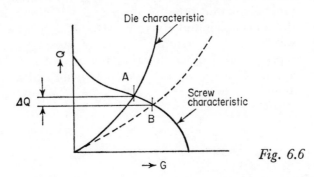

Fig. 6.6

The results from these calculations generally give answers which are experimentally verifiable with an error of around 10 per cent. This represents a vast improvement compared to results obtained by neglecting the flow component v_x.

Temperature effects in dies due to shear heating should really also be allowed for since they will affect the mean viscosity across the die profile. This has been attempted by Gee and Lyon [8] and by Griffiths [6].

More complicated cases of die design have also been analysed. Pearson [9, 10, 11] dealt with the thickness variations which arise from spider legs in tube dies and from the cross head in wire coating dies, reducing the problem to a two dimensional one on the basis of a shallow-depth wide channel.

7 Rheology of solid state behaviour—small deformations

Glass transition and crystalline melting point in amorphous and crystalline polymers. Viscoelastic stress response. Voigt, Maxwell and composite models. Creep and stress relaxation in static and dynamic situations. Generalised models. Distribution functions of retardation and relaxation times on linear and logarithmic time scales. Creep curves and isochronous stress/strain curves in the construction of three-dimensional response surfaces. Inertial effects leading to a dynamic situation. Mechanical-electrical analogies. The modulus of elasticity in product design (examples).

AN amorphous polymer may be rigid and glassy. This condition is due to the fact that the flow units are 'frozen' and locked. Alternatively it may be rubbery owing to the fact that the flow units are capable of movement with respect to one another when an external stress is applied. One and the same polymer can exhibit both conditions. Which response predominates depends on the environmental conditions of which temperature is the most obviously variable parameter. The transition can therefore be characterised by a temperature, the glass transition temperature Tg, which represents a fairly well defined reference point for glassy or rubbery behaviour under static conditions. When Tg is well below normal ambient conditions a polymer is regarded as a rubber. Since the rubbery condition is ultimately dependent on a sufficiency of free volume in which the flow units have some measure of freedom of movement, it is clear that any feature which results in the 'fluffing up' of main chains will increase free volume, reduce density and so depress Tg.

Some polymers consist essentially of amorphous material and such pseudocrystalline order as may exist can be ascribed to a fortuitous degree of order in essentially random entanglements. Others are highly ordered and may show crystallinity to an extent of 50 to about 95 per cent. Where the amorphous fraction is relatively small the glass transition will cease to be the dominant datum point for solid state behaviour, although the existence of *some* amorphous material may still allow a glass transition to be detected. It is clear, however, that the free volume necessary to allow the flow units of a crystalline polymer sufficient space to slide over one another implies their liberation from the restrictions of the crystalline structure itself. This necessitates heating to the crystalline melting point T_m, but since at this point the melt region is entered, one can hardly regard this as a true rubbery condition, even though the melt can sometimes be so viscous as to simulate a rubbery appearance when the material is manipulated. Thus:

(*a*) An amorphous polymer has a glass transition point above which the material becomes rubbery. On further heating it will gradually reduce in viscosity until it becomes a melt suitable for extrusion and moulding. The rubbery condition is used in sheet forming processes and it is no coincidence that the most important sheet forming polymers (acrylics, polystyrene, cellulose acetate and vinyl copolymers) are all amorphous.

(*b*) Highly crystalline polymers can only be rubbery in the sense that their melt viscosities happen to be high when the crystalline structure disappears at T_m. If the forces which maintain the crystal are weak (as in polythene or polypropylene), then T_m is low and the relatively cool melt will have a high melt viscosity. In this pseudo-rubbery state it is possible to use sheet forming techniques, but these require a high degree of control and some machinery modification. If, on the other hand, the forces which maintain the crystal structure are very strong, as for example in Nylon 66, then the high temperature needed to reach T_m will also cause the melt to be very fluid as soon as the melting point is reached, so that neither sheet forming, nor extrusion (except for special profiles such as fibre, monofilament, strip or film) are possible by ordinary techniques.

It may well be asked what significance the above has to solid state behaviour. The significance in fact is this: unless the concepts of Tg (associated with the amorphous phase) and T_m (associated with the crystalline phase) are fully appreciated it is difficult to refer sensibly to phenomena which represent the upper temperature limit of solid state behaviour and it is this which largely governs the choice of plastics for useful mechanical serviceability.

If a stress is applied to a polymer it can respond in three different ways:

1. It can show a rapid elastic response characterised by a high modulus which corresponds to bond stretching and the deformation of bond angles in the chain. This applies to both amorphous polymers below Tg and to crystalline polymers.

2. It can show viscous flow characterised by a low modulus which corresponds to the irreversible slipping of flow units past one another. This becomes possible in amorphous polymers above Tg and generally in all polymers in the melt.

3. It can show rubber elasticity characterised by a low modulus and by largely *reversible* slippage of flow units resulting in an elongation which may amount to several hundred per cent. As the flow units straighten out from their randomly coiled configuration they orientate themselves along the stress axis. This process is retarded by the internal frictional (viscous) forces of the specimen.

These response mechanisms may be schematised as in Fig. 7.1. At low temperatures the viscosities of the dashpots η_2 and η_3 are high and the polymer is only capable of a rapid elastic response defined by the spring modulus G_1.

At higher temperatures the viscosity of dashpot η_2 will become apparent and with it the second spring modulus G_2 and if this factor is dominant, then the polymer will be soft and rubbery rather than rigid. At still higher temperatures when the viscosity η_2 is low, the stress is quickly transmitted to the second dashpot of viscosity η_3 and when this becomes the dominant response, then we get true irreversible flow and little elasticity, since a restoring force in the shape of a spring in parallel with that dashpot is absent.

In practice all three processes come into operation, although one or other may be of major importance. This applies throughout the fluid (i.e. solid *and* liquid) state. The mixed response is termed 'viscoelasticity' and it is somewhat complicated to separate the behaviour into its viscous and elastic components. At any one temperature the extent to which each mechanism operates depends on the *rate* of stress application. Viscosity cannot show up under impact stress conditions, but under prolonged loading viscosity will tend to dominate and manifest itself in creep.

The spring and dashpot model, although useful as a conceptualising device, is rather crude and oversimplified. It suffers from two major limitations:

1. The dashpots suggest Newtonian flow and we know that Newtonian flow in polymers occurs only at very low shear stresses (the initial Newtonian region).

2. It applies only to amorphous polymers.

If it is accepted that we shall concern ourselves principally with amorphous polymers then we must also accept the further limitation of applying deformations which are small enough to permit the assumption that we are still operating within the linear viscoelastic region. In the solid state it is very difficult to be certain that any given small stress is small enough to fall within that region, so that conclusions as to the mechanical properties of materials based on linear viscoelastic

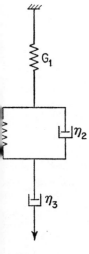

Fig. 7.1

theory are always suspect. However, an analysis based on classical linear viscoelastic theory is a useful starting point.

Models do not add anything new to the knowledge of viscoelasticity. Everything that can be derived from them is also available from phenomenological data. They were invented last century when polymers were unknown or unrecognised as such. But they give an easy picture of the significance

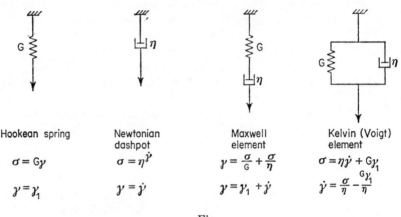

Fig. 7.2

of internal parameters of state. These parameters of state are represented by springs and dashpots. A spring stores energy on deformation; a dashpot dissipates energy on deformation. The internal parameters of *real* substances indicate to what extent the substances store and dissipate the energy applied by an external stress. The simplest mechanical models are shown in Fig. 7.2.

In the legend to Fig. 7.2, and in the following, the symbols used have the following meaning:

$$\sigma = \text{stress}$$
$$\gamma = \text{total strain}$$
$$\gamma_1 = \text{strain on spring}$$
$$\dot{\gamma} = \text{strain rate}$$
$$\eta = \text{(Newtonian) viscosity}$$
$$G = \text{spring modulus}$$

The reason for using a new symbol for stress in the solid state (σ in place of τ which was used for the liquid state) is that we want the symbol τ to denote a different function as will become evident presently. Strictly speaking, the above models should also include an inertial element (Fig. 7.3a for example). This may be omitted in large deformations of plastics since damping is substantial, but in small deformations, where damping is less, the result of inertial effects may be sinusoidal oscillations leading to Debye terms (see Appendix, p. 205ff). Fig. 7.3b shows this more clearly because the dashpot is omitted. When a force is applied to the inertial Hookean spring in Fig. 7.3b, it is obvious that on releasing the force recovery will produce a yoyo-like movement which in the absence of damping will produce a sinusoidal time/deformation curve with con-

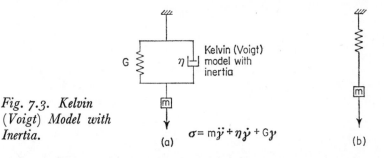

Kelvin (Voigt) model with inertia

Fig. 7.3. Kelvin (Voigt) Model with Inertia.

(a)

$$\sigma = m\ddot{y} + \eta\dot{y} + Gy$$

(b)

stant amplitude. The difference between Maxwell relaxation and Kelvin retardation models is as follows:

When springs and dashpots are arranged in series (Maxwell) they each bear the entire stress, but their deformations are additive.

When springs and dashpots are arranged in parallel (Kelvin) they each suffer the same deformation, but they divide the stress amongst themselves.

In order that the physical meaning of η/G for a Maxwell and Kelvin arrangement should not be confused it is denoted as the relaxation and retardation time respectively. The symbol τ is used whenever the two cannot be confused, but where the risk of confusion exists we are using τ_J for retardation time and τ_R for relaxation time.

The relaxation time is obtainable from a stress relaxation experiment at constant strain.

The retardation time is obtainable either from a creep experiment at constant stress or from a recovery experiment following a creep experiment on removal of the stress.

The Kelvin element is a good example of a viscoelastic body. If the model is stressed, part of the energy is stored in the spring whilst the remainder is dissipated in the dashpot and causes the deformation to be retarded. On releasing the stress recovery will occur due to the elasticity of the spring but is again retarded by the dashpot. The time needed for the system to achieve equilibrium is known as the *retardation time* τ_J.

Under a constant stress the Kelvin element shows creep, the deformation obeying the equation

$$J = J_0(1 - e^{-t/\tau_J})$$

where J = the creep compliance at time t, and
J_0 = the creep compliance at the time of stress application.

The compliances are the reciprocals of the moduli ($J = 1/G$).

If the applied stress is cyclic ('dynamic' stressing—see Chapter 9) then the storage and loss compliances are given by the Debye equations:

Storage:

$$J'(\omega) = J_0 \frac{1}{1 + \omega^2\tau_J^2}$$

Loss:

$$J''(\omega) = J_0 \frac{\omega\tau_J}{1 + \omega^2\tau_J^2}$$

where the frequency ω has the dimensions of reciprocal time.

It is clear that the Kelvin element is suitable for representing the retardation of a constant stress. But if we wish to investigate the relaxation of a constant deformation then the Kelvin element is not suitable and the Maxwell element has to be considered instead.

If a Maxwell element is deformed to a constant strain at zero time only the spring is stretched at the instance of stress application and the entire deformation energy is stored energy. But with the passing of time the dashpot flows under the influence of the restoring forces of the spring which thus doles out its stored energy to the dashpot for dissipation. The stress will relax and approach zero monotonically. As will be shown presently, this monotonic decay obeys the exponential equation

$$G(t) = G_0 \, e^{-t/\tau_R}$$

where τ_R again has the units of time and represents the 'relaxation time'. It is clear that when $\tau_R = t$ the modulus will have decayed to one-eth its original value. We can obviously also write $\sigma(t) = \sigma_0 \, e^{-t/\tau_R}$, because $G = \sigma/\gamma$ and is the same for any value of t under the arrangement whereby the strain is to remain constant.

In the case of dynamic (cyclic) strains—see Chapter 9—the analogous Debye equations for the storage and loss moduli are:

Storage:

$$G'(\omega) = G_0 \frac{\omega^2\tau_R^2}{1 + \omega^2\tau_R^2}$$

Loss:

$$G''(\omega) = G_0 \frac{\omega\tau_R}{1 + \omega^2\tau_R^2}$$

One of the advantages of dynamic tests lies in the fact that storage and loss moduli and compliances can be readily separated. This becomes apparent when the Debye equations are used in plotting

$$\ln\frac{t}{\tau_J} \quad \text{and} \quad \ln\frac{t}{\tau_R}$$

against J and G respectively (Fig. 7.4). Separate curves are obtained for the storage and loss modulus. The sum of these

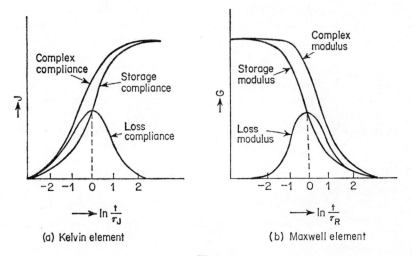

(a) Kelvin element

(b) Maxwell element

Fig. 7.4

two curves represents the complex dynamic compliance (or modulus) and the hysteresis loop between the complex and storage compliance (or modulus) therefore represents the 'hysteresis loss'.

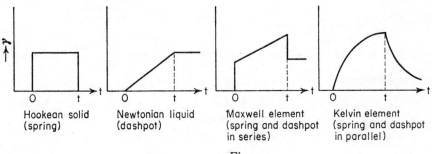

Hookean solid (spring)

Newtonian liquid (dashpot)

Maxwell element (spring and dashpot in series)

Kelvin element (spring and dashpot in parallel)

Fig. 7.5

Note that the two parts (*a*) and (*b*) of Fig. 7.4 are mirror images of each other.

Before we leave these simplest models (one and two parameters) let us just see what happens to the stress as a function of time when a constant strain is applied at zero time and released at time *t* (Fig. 7.5).

Looking back at Fig. 7.2, which includes the mathematical statement of the stress and strain relations for each type, it is clear that these cannot be used for the Maxwell and Kelvin elements as they stand, because the flow deformation which is part of the deformation of both depends not only on the stress but also on the duration of that stress. One must therefore first write down an expression giving the *rate* of change of deformation in the case of the Maxwell element. This will therefore be a differential equation. The flow displacement γ_2 is already a time derivative ($\dot{\gamma}$) by definition:

Equation 7.1
$$\dot{\gamma} = \frac{d\gamma_2}{dt} = \frac{\sigma}{\eta} \text{ (Newton's law)}$$

but

Equation 7.2
$$\gamma_1 = \frac{\sigma}{G} \text{ (Hooke's law)}$$

whence

Equation 7.3
$$\frac{d\gamma_1}{dt} = \frac{1}{G}\frac{d\sigma}{dt}$$

Summing the two derivatives we can write the total **deformation** rate as:

Equation 7.4
$$\frac{d\gamma}{dt} = \frac{d\gamma_1}{dt} + \frac{d\gamma_2}{dt} = \frac{1}{G}\frac{d\sigma}{dt} + \frac{\sigma}{\eta}$$

This fundamental differential equation determines the mechanical response of a material to a constant shear stress or strain. Obviously, at constant strain $d\gamma/dt = 0$ and the differential equation reduces to

Equation 7.5
$$\frac{1}{G}\frac{d\sigma}{dt} + \frac{\sigma}{\eta} = 0$$

which can then be integrated* to give

Equation 7.6
$$\sigma = \sigma_0\, e^{-Gt/\eta}$$

the law of Maxwellian decay in which the ratio η/G is the relaxation time τ_R. Since γ is constant, $\sigma/\sigma_0 = G/G_0$, and the law of Maxwellian decay can also be written:

* See Appendix.

Equation 7.7
$$G = G_0 \, e^{-t/\tau_R}$$

The behaviour of a Maxwell body in tension is similar to that in shear and the relaxation times will be the same, provided that the material is incompressible, because the tensile modulus and viscosities can then be assumed to be $3G$ and 3η respectively and their ratio is again τ_R.

Again looking back at Fig. 7.2, the basic differential equation for the Kelvin element can be made apparent from

$$\sigma = \eta \dot{\gamma} + G\gamma_1$$

by rewriting it

Equation 7.8
$$\frac{d\sigma}{dt} = \eta \frac{d\gamma_2}{dt} + G \frac{d\gamma_1}{dt}$$

which at constant stress can be set to zero and integrated,* giving

$$\gamma = \frac{\sigma}{G} \frac{1}{1 - e^{-Gt/\eta}}$$

where η/G is τ_J, the retardation time so that

Equation 7.9
$$\gamma = \frac{\sigma}{G} \frac{1}{1 - e^{-t/\tau_J}}$$

Upon removal of the stress the sample slowly returns to its original shape where $\gamma = 0$, following the experimental function:

Equation 7.10
$$\gamma = \gamma_0 \, e^{-t/\tau_J}$$

Thus, at constant stress, a retarded elastic specimen relaxes monotonically at a rate which is determined by its retardation time, just as at constant deformation a Maxwell body monotonically relaxes its stress at a rate which is determined by its relaxation time.

The relative importance of the elasticity/flow mechanism of response thus depends not only on G and η, but also on the experimental time scale t.

Thus, in a Maxwell model $(\sigma = \sigma_0 \, e^{-t/\tau_J})$, if t is large compared with the relaxation time τ_R, σ will approach zero. If it is small compared with τ_R, σ will approach σ_0. The same will apply exactly to the Kelvin model $(\gamma = \gamma_0 \, e^{-t/\tau_J})$ in a recovery experiment, taking γ_0 as the maximum deformation (at zero time) and substituting the appropriate symbols in the first statement.

Maxwellian decay can be plotted against t/τ_R in several ways. If log σ/σ_0 is plotted vs t/τ_R we get a straight line, as expected from Equation 7.7. But if instead we plot σ/σ_0 vs log t/τ_R we

* See Appendix.

obtain a curve of characteristic sigmoid shape, where a change in relaxation time τ_R does not alter the shape but merely displaces it along the time axis (Fig. 7.6). The curve is therefore universal for all materials which obey the Maxwell law of relaxation. The relaxation time is, of course, a function of free volume and so varies with temperature in any one material. It is therefore possible to obtain the complete master curve by carrying out the relaxation experiment at various temperatures to obtain the corresponding relaxation times and so get various pieces of curve which need only be shifted along the abscissa until all the pieces fit together. The degree of shifting depends on the relaxation time. This makes it possible to

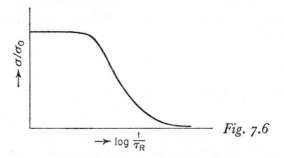

Fig. 7.6

predict the long term stress relaxation behaviour of a material from studies of its short term behaviour at various temperatures. This is known as the time/temperature *superposition principle*.

We have already seen that even for a rough approximation of model representation of plastics *at least* four parameters must be used. The representation shown (Fig. 7.1) is not the only possible one, however. It resembles a Kelvin type because it has a spring and dashpot in parallel. Another possible representation is shown in Fig. 7.7(*b*) which resembles the Maxwell type since it has two Maxwell elements in parallel. Since the relaxation and retardation times are ratios of a spring modulus and a dashpot viscosity ($\tau = \eta/G$) we can represent each viscosity (apart from that of the lone dashpot in the Kelvin composite) as a product $G\tau$. Both the Kelvin and Maxwell composites are equally representative—indeed they are mechanically equivalent. The Kelvin composite is preferred for calculating the time-dependent strain at constant stress (creep) whilst the Maxwell composite is preferred for calculating the time-dependent stress at constant strain (relaxation). If the parameters $(G_1)_R$, $(G_2)_R$, $(\tau_R)_1$, $(\tau_R)_2$ of the Maxwell composite and $(G_0)_J$ of the Kelvin composite are known, then the retardation time τ_J of the Kelvin composite can be calculated from the relationships:

$$(G_1)_R = (G_0)_J \frac{\dfrac{1}{(\tau_R)_1} - \dfrac{1}{\tau_J}}{\dfrac{1}{(\tau_R)_1} - \dfrac{1}{(\tau_R)_2}};$$

Equations 7.11

$$(G_2)_R = (G_0)_J \frac{\dfrac{1}{(\tau_R)_2} - \dfrac{1}{\tau_J}}{\dfrac{1}{(\tau_R)_1} - \dfrac{1}{(\tau_R)_2}}$$

where $(\tau_R)_1$ and $(\tau_R)_2$ are relaxation times and τ_J is a retardation time.

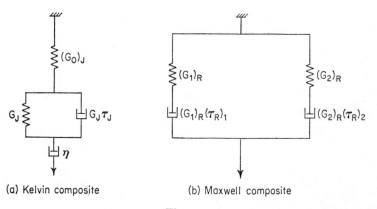

(a) Kelvin composite   (b) Maxwell composite

Fig. 7.7

We have dealt with simple models (Hooke, Newton, Maxwell, Kelvin) and four-parameter models of the Maxwell composite and Kelvin composite type as shown in Fig. 7.7. No further simplification of these is possible.

The simplification implicit in four-parameter models must now be adjusted in order to represent the actual behaviour of plastics more accurately: It just is not true that plastics are fully characterised by one or two relaxation or retardation times. The large numbers of possible configurations which are available to flow units must be reflected in a corresponding number of relaxation or retardation times. Some of the configurations will be more probable than others and in general the relaxation or retardation times will cluster around a central value in a Gaussian manner. The first step in arriving at a distribution function lies in constructing generalised Maxwell and Kelvin composites as shown in Fig. 7.8 below.

The generalised Kelvin model has an instantaneous elasticity parameter G_0, a Newtonian flow parameter η and n different Kelvin elements, each with its particular retardation time

(τ_J), and spring and dashpot constant; there are thus $(2n + 2)$ parameters present. The creep behaviour is given by the equation:

Equation 7.12
$$J = J_0 + \sum_{i=1}^{n} J_i (1 - e^{-t/(\tau_J)_i}) + t/\eta$$

The generalised Maxwell model consists of $(n + 1)$ Maxwell elements arranged in parallel, each with its particular relaxation time $(\tau_R)_i$ and spring and dashpot constant. Of the $(n + 1)$ relaxation times n correspond to the retardation times

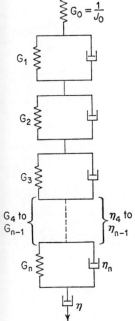

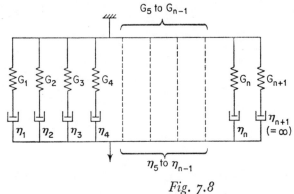

Fig. 7.8

Note. The dashpot $\eta_n + 1$ must have infinity viscosity, so that the spring G_{n+1} ensures that a finite stress remains after infinity time.

in the generalised Kelvin model and the additional one represents the flow in the generalised Kelvin model. Again, $(2n + 2)$ spring and dashpot parameters are present. The stress relaxation behaviour is given by the equation:

Equation 7.13
$$G = G_\infty + \sum_{i=1}^{n+1} G_i\, e^{-t/\tau_R)_i}$$

The next step is to extend the generalised Kelvin and Maxwell models from a finite number $(2n + 2)$ of springs and dashpots to an infinite number of springs and dashpots, so that their characteristic parameters vary continuously with their retardation or relaxation time. If this is done the result is, in each case, an infinite network characterised by a continuous function of a single independent variable, namely:

1. The distribution function of retardation times (retardation spectrum) for the continuous Kelvin model, and

2. The distribution function of relaxation times (relaxation spectrum) for the continuous Maxwell model.

The distribution functions can be fitted from experimental data after plotting the compliance or modulus against

$$\ln \frac{t}{\tau_J} \quad \text{or} \quad \ln \frac{t}{\tau_R}$$

for retardation and relaxation respectively. The resultant plot is shown in Fig. 7.9 below. The plot also includes the theoretical curve obtained from the Kelvin model assuming a single retardation time which is numerically equal to the mean retardation time of the continuous Kelvin model.

The obvious feature of Fig. 7.9 is the flattened experimental curve which renders the point of inversion much less incisive than the curve calculated on the assumption of a single retardation time. We shall return to this in Chapter 9.

Suffice it to say, that Equations 7.12 and 7.13, for creep and stress relaxation of generalised but still finite Kelvin and Maxwell models respectively, can be readily modified. All that is

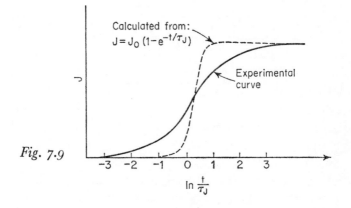

Fig. 7.9

necessary is to make the corresponding concomitant notational changes as set out in the table on the following page:

The derivations of Equations 7.14 and 7.15 are initially due to Wiechert [3] and Smekal [4] respectively and were subsequently reappraised by a number of workers.

G_∞, the integration constant in Equation 7.15 is the value of the modulus after infinity time. It represents the elasticity of the one spring which is connected to a dashpot of infinite viscosity (see note to Fig. 7.8) in the infinite model just as in the finite generalised Maxwell model. The spring must be included to ensure that a finite stress remains after infinity time.

J_0 in Equation 7.14 represents the instantaneous compliance, the integral term represents the viscoelastic deformation and t/η represents the flow (irreversible deformation) which occurs in creep.

| KELVIN | MAXWELL |

$$= J_0 + \sum_{i=1}^{n} J_i(\text{1} - e^{-t/\tau_J}) + \frac{t}{\eta} \quad \begin{bmatrix} \text{Equations for} \\ \text{generalised finite} \\ \text{models} \end{bmatrix} \quad G = G_0 + \sum_{i=1}^{n+1} G_i\, e^{-t/\tau_R}$$

the retardation times of the individual units $f(\tau_J)\, d\tau_J$ will be between zero and infinity.

the relaxation times of the individual unit $g(\tau_R)\, d\tau_R$ will be between zero and infinity.

Replacement of the summation sign $\sum_{i=1}^{n} J_i$ by the integral $\int_0^\infty f(\tau_J)$

Replacement of the summation sign $\sum_{i=1}^{n+1} G_i$ by the integral $\int_0^\infty g(\tau_R)$

The retardation spectrum $f(\tau_J)$ gives the contribution (to the equilibrium compliance of viscoelastic deformation) per unit interval of the time scale of the retardation processes with retardation times around τ_J.

$\begin{bmatrix} \text{Alfrey's definition} \\ \text{[2] of the functions} \\ f(\tau) \text{ and } g(\tau) \end{bmatrix}$

The relaxation spectrum $g(\tau_R)$ gives the contribution (to the equilibrium elastic modulus) per unit interval of the time scale of the relaxation processes with relaxation times around τ_R.

$$J = J_0 + \int_0^\infty f(\tau_J)(\text{1} - e^{-t/\tau_J}) \\ \times\, d\tau_J + \frac{t}{\eta}$$
Equation 7.14

$\begin{bmatrix} \text{Equations for} \\ \text{infinite models} \end{bmatrix}$

$$G = G\infty + \int_0^\infty g(\tau_R)(e^{-t/\tau_R})\, d\tau_R$$
Equation 7.15

The relaxation and retardation spectra $g(\tau_R)$ and $f(\tau_J)$ are appropriate for plots in which time is represented linearly. However, it is advantageous to use a logarithmic time scale for two reasons:

1. The time dependence of the processes becomes more pronounced.

2. Extrapolation to long time behaviour becomes both easier and more exact.

The conversion of the linear spectra is easily carried out by manipulation of the integral terms of Equations 7.14 and 7.15. We know that $dx/x = d \ln x$ and analogously $d\tau/\tau = d \ln \tau$, $d\tau = \tau\, d \ln \tau$. So, all we have to do to convert the linear equations for G and J is to multiply and divide the integral term by τ_R and τ_J respectively, e.g.:

$$\int_0^\infty f(\tau_J)(\text{1} - e^{-t/\tau_J})\, d\tau_J = \int_0^\infty \tau_J f(\tau_J)(\text{1} - e^{-t/\tau_J})\, \frac{d\tau_J}{\tau_J}$$

Writing:

$$
\begin{cases}
L(\ln \tau_J) \text{ for } \tau_J f(\tau_J) \quad \text{and} \quad \mathrm{d}\ln\tau_J \text{ for } \dfrac{\mathrm{d}\tau_J}{\tau_J} \\[3mm]
H(\ln \tau_R) \text{ for } \tau_R g(\tau_R) \quad \text{and} \quad \mathrm{d}\ln\tau_R \text{ for } \dfrac{\mathrm{d}\tau_R}{\tau_R}
\end{cases}
$$

we then have the integrals as the convenient expressions

$$
\int_{-\infty}^{+\infty} L(\ln\tau_J)(1 - e^{-t/\tau_J})\mathrm{d}\ln\tau_J
$$

and

$$
\int_{-\infty}^{+\infty} H(\ln\tau_R)\, e^{-t/\tau_R}\mathrm{d}\ln\tau_R
$$

$L(\ln\tau_J)$ still has the dimensions of a compliance and represents the contribution to creep of those retardation processes which have characteristic retardation times between $\ln\tau_J$ and $(\ln\tau_J + \mathrm{d}\ln\tau_J)$.

$H(\ln\tau_R)$ similarly still has the dimensions of a modulus and represents the contribution to stress relaxation of those relaxation processes which have characteristic relaxation times between $\ln\tau_R$ and $(\ln\tau_R + \mathrm{d}\ln\tau_R)$.

Substituting these modifications of the integral terms into Equations 7.14 and 7.15 we finally obtain:

Equation 7.16
$$
J = J_0 + \int_{-\infty}^{+\infty} L(\ln\tau_J)(1 - e^{-t/\tau_J})\,\mathrm{d}\ln\tau_J + \frac{t}{\eta}
$$

and

Equation 7.17
$$
G = G_\infty + \int_{-\infty}^{+\infty} H(\ln\tau_R)\, e^{-t/\tau_R}\,\mathrm{d}\ln\tau_R
$$

The instantaneous compliance J_0 and the equilibrium modulus G_∞ are obtained by setting $t = 0$, when the exponential terms disappear.

It will be noted that the integrals involving logarithmic time go from $-\infty$ to $+\infty$ whilst the integrals applying to linear time go from zero to $+\infty$. The reason for this is obvious: $\ln 0 = -\infty$.

A knowledge of L and H allows one to calculate all the visco-elastic functions. Strictly speaking, since the two are related [5], only one of them is necessary, but their relationship is cumbersome and one can carry out the necessary conversions by other means. This subject has been dealt with by a number of workers, for example: Passaglia and Knox in Baer: *Engineering design for plastics*, Ch. 3, p. 162; Andrew, *Ind. Eng. Chem.*,

1952, **44**, 708; Leaderman, in Eirich: *Rheology*, Vol. III; Ferry, *Viscoelastic properties of polymers*, Wiley, 1961.

Two of the most useful equations relating the functions are:

Equation 7.18
$$H(\ln \tau) = -\frac{dG}{d \ln t} + \frac{d^2G}{(d \ln t)^2}$$

where $t = 2\tau_R$ (Andrew), and analogously $L(\ln \tau_J)$ from J.

Equation 7.19
$$G = \frac{\sin m\pi}{m\pi J}$$

where m is the slope of the log J vs log t plot (Leaderman).

Equation 7.18 shows how the relaxation spectrum can be calculated from the relaxation modulus G; all that one requires is the first and second derivative of a plot of G vs ln t.

Equation 7.19 shows how the functions G and J can be calculated from one another.

An exceedingly concise and elegant way of arriving at Equations 7.17 and 7.18 has been given in a paper by Turner [6] who uses as his starting point the principal feature of the theory of linear viscoelasticity as it has been developed over the last fifty years, namely the assumption that stress and strain are related through a linear differential equation:

Equation 7.20
$$a_n \frac{\partial^n \sigma}{\partial t^n} + a_{n-1} \frac{\partial^{n-1} \sigma}{\partial t^{n-1}} + \ldots + a_0\sigma = b_m \frac{\partial^m \gamma}{\partial t^m}$$
$$+ b_{m-1} \frac{\partial^{m-1} \gamma}{\partial t_{m-1}} + \ldots + b_0\gamma$$

where $a_n \ldots a_0$ and $b_m \ldots b_0$ are material constants.

If this general equation is to be made rigorous one would have to transpose it into tensor notation. This would liberate one from the restrictive assumption that all the components of the stress or strain tensor are isotropic, but would also lead to formidable manipulating difficulties. In using Equation 7.20, then, as it stands, we are assuming:

1. that the stress/strain relationship is linear, and
2. that the stresses and strains are isotropic.

Both these assumptions are basically unjustified and can give no more than approximations and in plastics materials this is particularly evident. But Turner does no more than use these as a *starting point* for a transformation from unidimensional to three-dimensional linear viscoelasticity and then proceeds to consider the further complications which arise from non-linear viscoelasticity in plastics.

If we accept Equation 7.20 as correct and also accept the assumption stated above as serviceable approximations (however crude), then the equations for a Hookean body, a Maxwell element and a Voigt element follow immediately:

Hookean body: All constants a and b except a_0 and b_0 are zero. Equation 7.20 becomes:

Equation 7.21
$$a_0\sigma = b_0\gamma$$

Maxwell element: All constants a and b except a_0, a_1 and b_1 are zero.
Equation 7.20 becomes:

Equation 7.22
$$a_0\sigma + a_1\frac{\partial\sigma}{\partial t} = b_1\frac{\partial\gamma}{\partial t}$$

This is the spring and dashpot in series and applies to stress relaxation at constant strain.

Voigt element: All constants a and b except a_0, b_0 and b_1 are zero.
Equation 7.20 becomes:

Equation 7.23
$$a_0\sigma = b_0\gamma + b_1\frac{\partial\gamma}{\partial t}$$

This is the spring and dashpot in parallel and applies to strain retardation at constant stress.

Neither the Maxwell nor the Voigt element and therefore neither Equation 7.22 nor Equation 7.23 can represent both creep and stress relaxation simultaneously, but a combination of the equations—or a modification of Equation 7.20 in which all constants a and b except a_0, a_1 . b_0 and b_1 are zero—does so. The resulting Equation 7.24 expresses this in mathematical terms:

Equation 7.24
$$a_1\frac{\partial\sigma}{\partial t} + a_0\sigma = b_1\frac{\partial\gamma}{\partial t} + b_0\gamma$$

Equations 7.22, 7.23 and 7.24 have simple solutions involving the term $e^{-t/\tau}$, where τ is a function of the constants a and b.

If the strain is constant τ is τ_R, the relaxation time;

If the stress is constant τ is τ_J, the retardation time.

The approximate solution of these equations can be improved if the n and m terms in Equation 7.20 are *not* assumed to be negligibly small. The alternative method of representation involves large numbers of Maxwell elements in parallel or Voigt elements in series. These are the generalised models which have been presented before, with a *series* of relaxation or retardation times respectively.

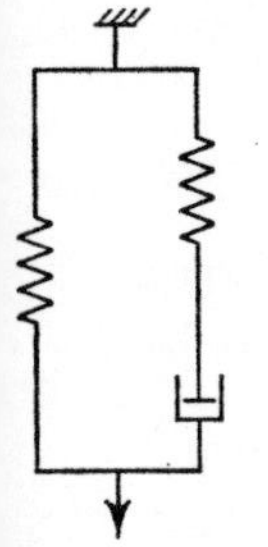

Equation 7.22, on solution, gives

del of eq^n (24)

Fig. 7.9(a)

Equation 7.25a $$\sigma = \sigma_0\,e^{-t/\tau_R}$$

which for a generalised model becomes

Equation 7.25b $$\sigma = \sum_{i=1}^{N} (\sigma_0)_i\,e^{-t/(\tau_R)}$$

and

Equation 7.25c $$G(t) = \frac{\sigma}{\gamma_0} = \sum_{i=1}^{N} \frac{(\sigma_0)_i}{\gamma_0}\,e^{-t/(\tau_R)_i}$$

where $G(t)$ is the time dependent relaxation modulus, or

$$G(t) = \frac{\text{total (time dependent) stress}}{\text{(constant) strain}}$$

Summing over the entire range of infinitely small increments of relaxation times, the summation is replaced by the integral:

Equation 7.25d $$G(t) = \int_0^{\infty} g(\tau_R)\,e^{-t/\tau_R}\,d\tau_R + G_{\infty}$$

where $g(\tau_R)$ is the distribution function of relaxation times (the relaxation spectrum) and G_{∞} is the integration constant (the value of the modulus at infinity time).

This, then, is the treatment of Maxwell relaxation. The treatment of Voigt retardation is precisely analogous:

First we solve Equation 7.23 to obtain

Equation 7.26a $$\gamma = \gamma_0(1 - e^{-t/\tau_J})$$

which for the generalised model becomes

Equation 7.26b $$\gamma = \sum_{i=1}^{N} (\gamma_0)_i(1 - e^{-t/\tau_J})$$

and

Equation 7.26c $$J(t) = \frac{\gamma}{\sigma_0} = \sum_{i=1}^{N} \frac{(\gamma_0)_i}{\sigma_0} (1 - e^{-t/\tau_J})$$

where $J(t)$ is the time-dependent creep compliance, or

$$J(t) = \frac{\text{total (time-dependent) strain}}{\text{(constant) stress}}$$

Summing over the entire range of infinitely small increments of retardation times, the summation sign is replaced by the integral:

Equation 7.26d $$J(t) = \int_0^{\infty} f(\tau_J)(1 - e^{-t/\tau_J}) + J_0 + \frac{t}{\eta}$$

where $f(\tau_J)$ is the distribution function of retardation times (the retardation spectrum) and J_θ is the integration constant (the value of the instantaneous compliance at zero time).

The term t/η is the flow term which becomes important when t is very large (since η is *always* very large) and which is neglected in relatively short-time experiments.

There remains only one more modification to be made: It is obviously preferable to plot $G(t)$ or $J(t)$ vs log time rather then vs linear time; the transformation of Equations 7.25d and 7.26d are readily accomplished in the manner already shown, giving respectively:

Equation 7.25e
$$G(t) = \int_{-\infty}^{+\infty} H(\ln \tau_R)\, e^{-t/\tau_R}\, \mathrm{d}\ln \tau_R + G_\infty$$

and

Equation 7.26e
$$J(t) = \int_{-\infty}^{+\infty} L(\ln \tau_J)(1 - e^{-t/\tau_J})\mathrm{d}\ln \tau_J + J_0 + \frac{t}{\eta}$$

Finally it should be mentioned that the *dynamic* parameters (storage and loss modulus, storage and loss compliance) can also be expressed in logarithmic distribution functions [7].

The dynamic moduli and compliances are, of course, a function of frequency which replaces the simple time function and becomes its analogue when transposing from static to dynamic stress/strain relationships. The function of the form $e^{-t/\tau}$ is replaced by the appropriate and characteristic Debye term of the form $1/1 + \omega^2\tau^2$. In the final result we obtain:

Equations 7.25f

Storage Modulus
$$G'(\omega) = \int_{-\infty}^{\infty} H(\ln \tau_R)\, \frac{\omega^2\tau_R^2}{1 + \omega^2\tau_R^2}\, \mathrm{d}\ln \tau_R + G_\infty$$

Loss Modulus
$$G''(\omega) = \int_{-\infty}^{\infty} L(\ln \tau_J)\, \frac{\omega\tau_R}{1 + \omega^2\tau_R^2}\, \mathrm{d}\ln \tau_R$$

Equations 7.26f

Storage Compliance
$$J'(\omega) = \int_{-\infty}^{\infty} L(\ln \tau_J)\, \frac{1}{1 + \omega^2\tau_J^2}\, \mathrm{d}\ln \tau_J + J_0$$

Loss Compliance
$$J''(\omega) = \int_{-\infty}^{\infty} L(\ln \tau_J)\, \frac{\omega\tau_J}{1 + \omega^2\tau_J^2}\, \mathrm{d}\ln \tau_J + \frac{1}{\omega\tau_J}$$

Why are Equations 7.25e and 7.26e so important? In S. Turner's words their principal importance lies in this:

It is characteristic of linear systems, i.e. those representable by a linear differential equation that if the response to a step input is known (e.g. from stress relaxation and creep experiments), then the response to any arbitrary input can be calculated from the superposition integral, a procedure often known as Boltzmann's Superposition Principle within the field of linear viscoelasticity [6].

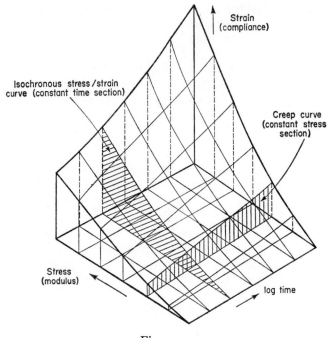

Fig. 7.10

The phenomenon of creep in thermoplastics has been very satisfactorily dealt with by Turner [8] who also gives ten references in his review. He points out that whilst stress-strain-time temperature relationships can be established in a number of ways, the creep experiment has two important advantages, namely:

1. It is simple;
2. It can be directly applied to service problems, especially with the more rigid plastics in which designers are becoming increasingly interested.

The importance of the creep function lies in the fact that extrapolation from a limited data range will give an acceptable account of creep behaviour over a much wider range of variables, even though extrapolation techniques are never entirely reliable. The viscosity term in the creep function (Equation 7.17) rarely occurs at small deformations, except with very low molecular weight polymers or at high temperatures.

By keeping the temperature constant the stress-strain-time surface can be produced for a limited number of constant stress (creep) and constant time (isochronous) sections as shown in Fig. 7.10. The solid model can be reduced to a contour map, each contour representing an iso-strain or stress relaxation

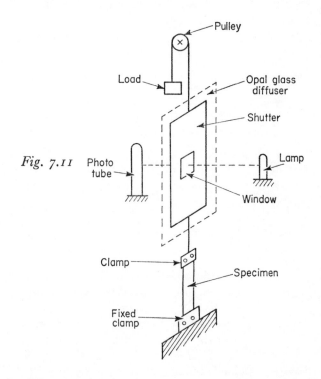

Fig. 7.11

curve. A number of contour maps each representing a different constant temperature in turn will complete the picture.

Apparatus for creep measurements is reviewed in a paper by Dunn, Mills and Turner [9] but special mention should be made of the elegant device described by Scherr and Palm [10]. It consists of an open rectangular frame with a fixed specimen clamp at the bottom end and a ballbearing pulley at the top. The applied load pulls the specimen upwards and the unit can be conveniently put into a constant temperature chamber (Fig. 7.11). The sensing part of the apparatus is a photoelastic transducer. A phototube is mounted on a steel channel frame and receives an exciting radiation from a projection lamp located opposite after passage through an opal glass diffuser. The phototube detects the creep (or recovery) of the specimen since its cathode surface will become increasingly shadowed by the upward movement of the window in the shutter which is directly attached to the free end of the specimen. The phototube signal is amplified and automatically recorded. The creep (in dynes cm^{-2}) is plotted vs ln time and was found to give excellent agreement with cathetometer results. The advantages of the apparatus are that it does not require any operator attention, that a temperature range from -40 to $150°C$ can be covered at $0.5°$ intervals and that the load is easily variable.

So far we have neglected the effects of inertia. This is reasonably justified in plastics in which only static stresses are

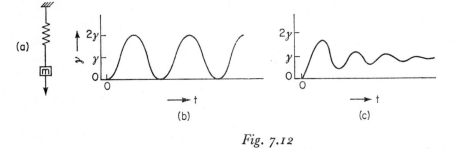

Fig. 7.12

considered because in these no acceleration of volume elements occurs. But the sudden application of a stress could result in oscillating responses, especially in an elastomer. We are here entering the border region between static and dynamic stresses and the dynamic behaviour of polymer depends on both the inertia and the elastic restoring forces. Supposing that a polymer specimen has a sudden stress applied to it at the two ends. At the moment of stress application the central part of the mass experiences no stress at all and it will take some finite time before elastic waves become established throughout the specimen. The time necessary to establish this resonance state will depend on the distance of separation of the points of stress application as well as the velocity of wave propagation. At first however a state of vibration is created which will be non-uniform throughout the specimen and which will also vary with time.

Let us first consider the simplest case of a spring, and lump the inertia together as a mass hanging from the bottom of the spring instead of distributing it uniformly along the spring (Fig. 7.12a). When a stress is now applied the strain is no longer given by Hooke's law $\sigma = G\gamma$ alone but includes an additional term of inertial acceleration, so that the equation becomes:

Equation 7.27
$$\sigma = m\frac{d^2\gamma}{dt^2} + G\gamma$$

m being an inertial mass which is proportional to the density.

If a stress is suddenly applied at zero time, then the strain will not immediately settle into the equilibrium deformation $\gamma = \sigma/G$, but will oscillate between zero and 2γ as shown in Fig. 7.12b. In actual fact, however, the oscillation is damped out by the internal viscosity of the material, so that the strain/time diagram takes the shape of Fig. 7.12c.

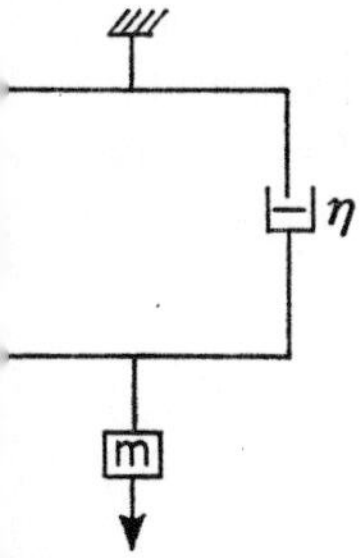

Let us now consider inertial mass to be added to a Kelvin element (Fig. 7.13). Obviously, the basic equation used before,

$$\sigma = \eta\frac{d\gamma}{dt} + G\gamma$$

Fig. 7.13

must again be modified by addition of the term of inertial acceleration to give:

Equation 7.28
$$\sigma = m\frac{d^2\gamma}{dt^2} + \eta\frac{\gamma}{dt} + G\gamma$$

When the damping out of oscillations as shown in Fig. 7.12c is great, then the specimen attains its equilibrium deformation very rapidly and the inertial character of elasticity may be ignored. This is generally the case with plastics when a constant stress is applied gradually. Not so, however, in cases where the stress is applied very rapidly (impact conditions) or where it varies cyclically. When is a stress applied sufficiently rapidly for inertial effects to become important? This is obviously a function of the viscoelastic parameters of the material and on reflection it becomes obvious that relaxation times are decisively involved. As far as cyclic stresses are involved, this will receive attention in Chapter 9. The mathematical solution of Equation 7.28 is given in the Appendix, Equation A1, etc., p. 200.

The sudden application of a stress by mechanical means is impossible—the forces are applied at the surface and must therefore set up waves which travel at finite rates. A constant stress distribution is only obtained when the initial waves are damped out. If this takes appreciable time one must be careful when interpreting stress responses in the initial stages of an experiment. The justification for introducing the subject of inertial and damped inertial elasticity which does not normally figure very importantly in static testing lies precisely in expressing this warning and explaining the reasons which underlie it.

Whilst mechanical stresses cannot be applied instantaneously and whilst they can only be transmitted from the surface, other types of stresses *can* be applied virtually instantaneously and uniformly throughout the entire volume of a test specimen. The most important type of stress which falls into the latter category is that which is due to the application of an electric field. This acts on such dipoles as may be present and distorts them like springs against the various opposing forces present. Mathematically, the differential equations which apply to mechanical systems apply equally to other force fields. Stambough [11] has drawn up the following table of analogies between mechanical and electrical stress systems:

Mechanical	Electrical
Load (gm mass)	Voltage
Strain	Current velocity
Behaviour of dashpot (viscosity η)	Behaviour of resistance (R)
Behaviour of spring (compliance J)	Behaviour of condenser (capacitance C)
Series connection of spring and dashpot (Maxwell)	Parallel connection of condenser and resistance
Parallel connection of spring and dashpot (Kelvin)	Series connection of condenser and resistance
Mechanical energy stored in springs	Electrical energy stored in capacitors
Mechanical energy dissipated in dashpots	Electrical energy dissipated in resistances
Calculation of G in a generalised Maxwell body (tedious)	Analysis of mechanical behaviour using electrical network systems (this is more easily variable to fit experimental curves)

The actual determination of relaxation spectra can be achieved from an evaluation of available relaxation and steady-state flow data [12].

Starting from the relation between the decaying stress $\sigma(t)$ and the steady-state shear rate $\dot{\gamma}$

$$\sigma(t) = \dot{\gamma} \int_{-\infty}^{\infty} H(\tau)\tau \, \mathrm{d}\ln\tau \, \mathrm{e}^{-t/\tau}$$

where $H(\tau)$ is the distribution function of relaxation times on a logarithmic time scale, it is necessary to invert this function so as to make $H(\tau)$ the subject. This can be done by approximation methods as shown by Ferry and co-workers [13, 14] and leads to a first approximation of:

$$H(\tau)\Big|_{\tau=t} = -\frac{1}{\dot{\gamma}t}\frac{d\sigma(t)}{d\ln t}$$

The logarithmic time scale spectrum $H(\tau)$ can also be evaluated from steady-state flow data as shown by Faucher [15] who has extended the non-Newtonian flow theory of Eyring and Ree [16]. Faucher, assuming a continuous distribution of flow units, rewrites the Eyring-Ree equation

$$\sigma(\dot{\gamma}) = \int_0^\infty G(\tau) \text{ arc sinh } (\dot{\gamma}\tau) \, d\tau$$

where $G(\tau)$ is the ordinary relaxation spectrum as defined before, namely

$$\tau G(\tau) = H(\tau)$$

Differentiating the Eyring-Ree equation, one obtains:

$$\frac{d\sigma}{d\dot{\gamma}} = \int_0^\infty \frac{\tau G(\tau) \, d\tau}{\sqrt{1 + \dot{\gamma}^2\tau^2}}$$

which can be transposed into two parts, viz:

$$\frac{d\sigma}{d\dot{\gamma}} = \int_0^{\tau=(1/\dot{\gamma})} \frac{\tau G(\tau) \, d\tau}{\sqrt{1 + \dot{\gamma}^2\tau^2}} + \int_{\tau=(1/\dot{\gamma})}^\infty \frac{\tau G(\tau) \, d\tau}{\sqrt{1 + \dot{\gamma}^2\tau^2}}$$

For small shear rates, i.e. $\dot{\gamma} \leqslant 1$ the first integral represents the main contribution, and, neglecting the second, one obtains:

$$G(\tau)\Big|_{\tau=(1/\dot{\gamma})} \cong -\dot{\gamma}^3 \frac{d^2\dot{\gamma}}{d\dot{\gamma}^2}$$

or

$$H(\tau)\Big|_{\tau=(1/\dot{\gamma})} \cong -\dot{\gamma}^2 \frac{d^2\dot{\gamma}}{d\dot{\gamma}^2}$$

On the other hand, if $\dot{\gamma} > 1$, the second integral provides the main contribution, which, as Faucher also shows, leads to

$$G(\tau)\Big|_{\tau=(1/\dot{\gamma})} \cong \dot{\gamma}^3 \frac{d^2\sigma}{d\dot{\gamma}^2} + \dot{\gamma}^2 \frac{d\sigma}{d\dot{\gamma}}$$

or

$$H(\tau)\Big|_{\tau=(1/\dot{\gamma})} \cong \dot{\gamma}^2 \frac{d^2\sigma}{d\dot{\gamma}^2} + \dot{\gamma} \frac{d\sigma}{d\dot{\gamma}}$$

Clearly, we have here the 1st and 2nd derivatives of the $\sigma/\dot{\gamma}$ curve which can be determined by graphical differentiation.

Experimentally, it is necessary to follow the stress relaxation of a melt (say in a cone/plate viscometer) after reaching, maintaining and cessation of a steady-state shear.

In their work on polystyrene and PMMA, Ajroldi, Gar-burglio and Pezzin [12] found that the equation

$$H(\tau)\Big|_{\tau=(1/\dot{\gamma})} \cong -\dot{\gamma}^2 \frac{\mathrm{d}^2\sigma}{\mathrm{d}\dot{\gamma}^2}$$

gave good agreement between experimental and calculated spectra *even for* $\dot{\gamma} > 1$ and that the alternative equation

$$H(\tau)\Big|_{\tau=(1/\dot{\gamma})} \cong \dot{\gamma}^2 \frac{\mathrm{d}^2\sigma}{\mathrm{d}\dot{\gamma}^2} + \dot{\gamma}\frac{\mathrm{d}\sigma}{\mathrm{d}\dot{\gamma}}$$

gave poor results in any case, evidently because the second integral of the equation

$$\frac{\mathrm{d}\sigma}{\mathrm{d}\dot{\gamma}} = \int_0^{\tau=(1/\dot{\gamma})} \cdots + \int_{\tau=(1/\dot{\gamma})}^{\infty} \cdots$$

approaches zero and equation

$$H(\tau)\Big|_{\tau=(1/\dot{\gamma})} \cong -\dot{\gamma}^2 \frac{\mathrm{d}^2\sigma}{\mathrm{d}\dot{\gamma}^2}$$

can be applied to *any* range of times or shear rates.

On the other hand, the Ree-Eyring equation is:

$$\eta_a = \sum \frac{x_n}{\alpha_n} \beta_n \frac{\text{arc sinh } (\beta_n\dot{\gamma})}{\beta_n\dot{\gamma}}$$

where β_n is proportional to the τ of the nth flow unit;

$\quad\quad x_n$ is the frictional area of the nth flow unit;

$\quad\quad \alpha_n$ is the ratio $\dfrac{\text{shear volume}}{2 \times \text{kinetic energy}}$ of the nth flow unit.

If one assumes β_n to be identical with the relaxation time τ and also assumes a continuous distribution of τ one obtains:

$$\eta_a = \int_0^{\infty} \tau G(\tau) \frac{\text{arc sinh } (\dot{\gamma})}{\dot{\gamma}\tau}\, \mathrm{d}\tau$$

If we plot (arc sinh x/x) vs log x it is seen that the point of maximum slope is at $\tau\dot{\gamma} = 2\cdot9$. Now,

$$\frac{\text{arc sinh } \dot{\gamma}\tau}{\dot{\gamma}\tau} \begin{cases} = 0 & \text{when} \quad \dot{\gamma}\tau < 2\cdot9,\ \text{and} \\ = 1 & \text{when} \quad \dot{\gamma}\tau > 2\cdot9 \end{cases}$$

and by substituting for the inverse hyperbolic sine function one obtains:

$$\eta_a \cong \int_0^{\tau=(2\cdot9/\dot{\gamma})} \tau G(\tau)\, \mathrm{d}\tau$$

from which, on differentiation,

$$H(\tau)\Big|_{\tau=(2\cdot9/\dot\gamma)} \cong -\frac{\dot\gamma^2}{2\cdot9}\frac{\mathrm{d}\eta_a}{\mathrm{d}\dot\gamma}$$

Now this equation can be directly applied to flow data and this affords a second means for obtaining $H(\tau)$ which agrees quite well with

$$H(\tau)\Big|_{\tau=(1/\dot\gamma)} \cong -\dot\gamma^2\frac{\mathrm{d}^2\sigma}{\mathrm{d}\dot\gamma^2}$$

both of which are really equivalent by inspection.

[$\log H(\tau)$ in dynes cm^{-2}]

One of the most interesting aspects which emerges from this, by the way, is the fact that a relaxation spectrum equation originally derived from solid state viscoelastic considerations yields valid results when applied directly to the liquid (melt) state of polymeric materials, a further powerful corroboration of the concept of the integrity of a generalised fluid state.

For the conclusion of this account nothing could be more suitable than examples which illustrate how a knowledge of moduli can be used to solve design problems. The two examples given below are taken from a recent paper whose author reference was unfortunately mislaid by the writer and who wishes to acknowledge his indebtedness by giving an undertaking that the regrettable omission will be remedied when possible.

Example 1

A cantilever 4 in. long is to hold 10 lbs under continuous load for one year. The maximum temperature of the ambient air is 40°C and the deflection of the beam must not exceed $\frac{1}{8}$ in. If a rectangular beam of acetal resin is to be used, the modulus has been found to be 150,000 psi. What are the required dimensions of the beam?

Solution

The standard formula for a stressed beam is:

$$\gamma = \frac{\sigma l^3}{3GI}$$

where γ = deflection = 0·125 in. (max.)
        σ = load = 10 lbs
        l = length of beam = 4 in.
        G = modulus = 150,000 psi
        I = moment of inertia

Hence $I = 0\cdot0114$ in.

For a rectangular cross section:

$$I = \frac{bd^3}{12}$$

where b is the width and d is the depth of the beam normal to the load.

If b is 0·25 in., then,

$$I = \frac{0·25d^3}{12} = 0·0114$$

whence $d^3 = 0·545$ and $d = 0·82$ in. Similarly, it is possible to plot d as a function of b and the customer can take his pick.

Example 2

Determine the bulge to be expected in an aerosol bottle with flat base and cylindrical walls which is to remain stored on a shelf for one year with an internal pressure of 100 psi. The thickness of the base is 0·15 in., the radius 0·75 in. Assuming that the equation for the deflection at the centre of a plate with supported edges is

$$\gamma = \frac{3Pr^4(5 - 4\mu - \mu^2)}{16G_yd^3}$$

where P = pressure in psi = 100

$\quad\quad R$ = radius = 0·75

$\quad\quad \mu$ = Poisson's ratio = 0·38 for the acetal resin intended for use

$\quad\quad G_y$ = apparent modulus at one year = 160,000 psi at 40°C

$\quad\quad d$ = thickness = 0·15

Hence $\gamma = 0·37$ in.

An actual bottle showed a bulge of 0·34 in. under these conditions. When similar calculations were made for a circular plate fixed at the edges, so that the internal diameter cannot be reduced, then the estimated bulge is 0·009 in. The actual figure is intermediate because the bottle walls behave in a manner intermediate between the two calculated cases. A bottle designed with an inward bulge of more than 0·037 in. will therefore not be likely to deform under pressure sufficiently to give a convex bottom which will rock.

8 Rheology applied to solid state behaviour—large deformations

The generalised fluid state. The generalised flow curve: analogies between solid and liquid stress responses. Conventional stress and true stress; the Considère construction for determining the onset of strain-softening and strain-hardening. Tough and brittle failure. The severity of test as the only criterion for tough or brittle failure in components. Methods of increasing the severity of test: temperature, time/frequency, composition of the complex stress system, previous stress history, and the ultimate dependence of all these on free volume; the effect of crosslinkage, molecular architecture, environmental chemical stresses, additives. Crack initiation and crack propagation. The limitations of materials tests in product design. 'Brittle points.' Uniaxial tension—a pure or a composite stress system? The importance of notches in modifying real complex stress systems. Isotropic and anisotropic stress. Summary of variables when performing materials tests. A realistic scheme for obtaining a reasonably comprehensive understanding of materials behaviour for design purposes.

It has been seen in Chapter 7 that one of the basic parameters of a material in the solid state is the modulus which has the same significance for that state as viscosity has for liquids. The main difference between an ideal solid and an ideal liquid is that the deformation of the former is a dimensionless magnitude whilst deformation in the latter can only be expressed as a function of time. Since neither ideal solids nor ideal liquids do in fact exist it is best to consider them as extreme poles of a more or less continuous fluid state.

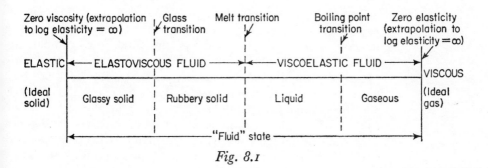

Fig. 8.1

The terms 'viscoelastic' and 'elastoviscous' merely indicate that the first part of the stress response characteristic predominates and normally this distinction is implied rather than formally expressed. For a rheologist it is essential that he should not allow himself to be mesmerised by the discontinuities of transitions of state. These transitions are, of course, of great importance because they produce abrupt changes in free volume and/or heat content, with all the thermodynamic

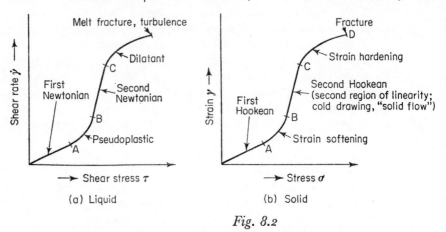

Fig. 8.2

implications involved, but this should not obscure the concept of an all-embracing generalised fluid state.

The most powerful argument in favour of this concept is the evident parallelism between stress/strain curves of solids and shear stress/shear rate curves of liquids when these are drawn between the limits of zero to infinity stress. This involves not only the linear deformations within the proportionality limit but also the non-linear deformations beyond. It is the latter which are of dominant interest in the design of solid components, just as it is for the processing of the liquid melts. This chapter is devoted to the entire viscoelastic response in the solid state, with particular emphasis on the non-linear region.

Just as, in the liquid state, we have a generalised flow curve as postulated in Chapter 1, so there exists a deformation curve in the solid state which shows the analogous component elements (Fig. 8.2).

Stress/strain curves are usually plotted with stress as the ordinate and the reverse method was used in Fig. 8.2 to emphasise the parallelism. Some workers prefer to plot shear stress as the ordinate for reasons of consistency; but it may also be argued, again on the grounds of consistency, that the *independent* variable should be plotted as the ordinate in both cases. This usually happens to be the shear rate in liquids and the stress in solids. Following the latter convention, Fig. 8.2b will appear mirrored across the 45° axis (Fig. 8.3a). Such a stress/strain curve is not *directly* obtainable from tensile measurements since the stress is expressed in force per unit area of the *original* specimen cross section and the specimen cross section obviously changes continuously during the experiment and so produces a curve of the type as shown in Fig. 8.3b. It is, however, possible to correct the actual stress/strain curve if

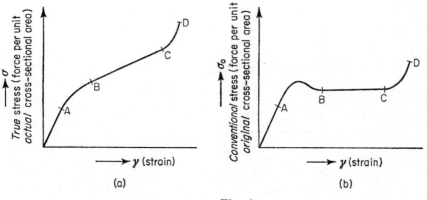

Fig. 8.3

the change in cross sectional area is determined during the experiment. If this is done and stress (in terms of force per actual cross sectional area) is replotted against strain, then a curve as shown in Fig. 8.3a is in fact obtained.

In the modulus considerations dealt with in Chapter 7 only the linear region of small scale deformations was treated. Modulus, like viscosity, has its proportionality limit set by the position of point A and to determine apparent moduli as the slope of the line drawn from the origin to a point beyond A is, strictly speaking, just as meaningless as to determine the apparent viscosity from a flow curve by the analogous procedure. Since the proportionality limit is, however, quite low in some materials, engineers sometimes go slightly beyond point A, so long as the apparent modulus is not less than 85 per cent of the initial modulus, and use the corresponding deformation in design calculations.

Whilst small scale deformations are characterised by the modulus which is constant up to the proportionality limit, the modulus changes at larger deformations and it is obviously essential to know the previous stress history of a component under test. No mouldings, extrusions, or calendered products are ever free from built-in stresses and strains unless they are carefully annealed and it is therefore obvious that processing conditions can cause great differences in products made from the same material. Assuming, however, that the stress/strain history slate of a component has been wiped clean by annealing and that it has thus been returned to its ground state, then it should be possible to lay down the conditions for strain softening, necking, cold drawing and strain hardening, as has been shown by Vincent [1].

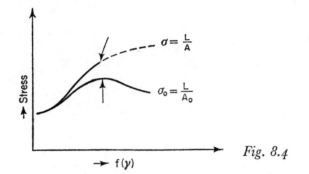

Fig. 8.4

Let the conventional stress σ_0 be defined by L/A_o, where L is the load and A_o is the original cross-sectional area; let the true stress $\sigma = L/A$, where A is the actual cross-sectional area at the time of applying load L.

Let l be the length of a tensile specimen under load L, l_o the original length and $l/l_o = R$, the 'elongation ratio'.

If σ_0 and σ are both plotted as a function of strain defined by

$$f(\gamma) = \frac{l - l_o}{l_o} = R - 1$$

then two stress/strain curves are obtained as shown in Fig. 8.4.

If σ_0 falls below the arrowed maximum then the specimen begins to neck. Assuming incompressibility and constant geometrical shape, the volume remains constant during the experiment and also remains definable by:

$$Al = A_o l_o$$

This can be written

$$\frac{A_o}{A} = \frac{l}{l_o}$$

which is identical with the 'elongation ratio' R.

Hence $A = A_o/R$, and substituting for A we get

$$\sigma = \frac{L}{A} = \frac{LR}{A_o} = \sigma_0 R$$

so that σ, the true stress, is readily accessible up to the point where necking starts. Necking is a shape discontinuity characterised by a change of the specimen boundary surfaces from the condition of parallelism with the direction of the force (Fig. 8.5).

Once necking has started, the relationship $\sigma = R\sigma_0$ can no longer be used, because the prismatic or cylindrical shape assumed in deriving the relationship is no longer in existence.

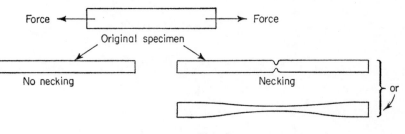

Fig. 8.5

If the curve for σ in Fig. 8.4 were to be extended, one would have to obtain the true stress from the force and the dimensions of the thinnest part of the neck (dotted continuation in Fig. 8.4).

Up to the point of necking (arrowed in Fig. 8.4) the relationship $\sigma_0 = \sigma/R$ holds. The point of necking is defined by the maximum of the σ_0 curve, where the slope of the tangent $d\sigma_0/dR$ is equal to zero. We can therefore differentiate σ_0 with respect to R and set to zero:

$$\frac{d\sigma_0}{dR} = \frac{1}{R}\frac{d\sigma}{dR} - \frac{\sigma}{R^2} = 0$$

whence

$$\frac{d\sigma}{dR} = \frac{\sigma}{R}$$

and this we also know to be equal to σ_0 (see before).

Therefore, if the true stress σ is plotted vs R, the point where necking starts is that point on the curve where the slope of the plot is equal to σ_0 (the conventional stress). This point may be found by drawing a tangent to the curve from the origin ($\sigma = 0$, $R = 0$) and this is known as 'Considère's construction' (Fig. 8.6).

The actual place where the neck forms is a place of weakness in the specimen which is always present because of the impossibility of making the specimen perfectly uniform along its entire length. The stress will therefore be greater at one spot than anywhere else. In the initial Hookean region of the stress/strain curve where the curve rises steeply this non-uniformity does not yet cause a neck to appear because the slight extra strain can be supported in the vulnerable spot without producing an excessive stress. But as the modulus decreases beyond the proportionality limit, additional stresses produce progressively greater strains and this will tend to make the vulnerable spot thinner than the rest of the specimen. Eventually the system becomes unstable and a neck is formed.

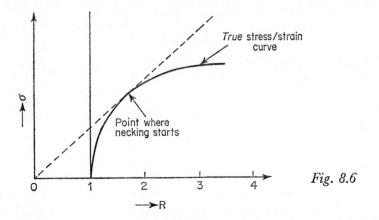

Fig. 8.6

Strain softening, having started at the point of departure from Hookean proportionality, culminates in producing maximum instability in the system at the point given by Considère's construction.

When a neck has been formed one of two things can happen on further application of stress:

(a) *Either* the neck becomes progressively thinner and eventually the specimen breaks;

(b) *Or* the neck may stabilise itself to a constant cross-sectional area as the shoulders travel along the specimen until they have 'consumed' all available undrawn material. During this process the nominal load remains constant. This is called *cold-drawing*.

The neck is finally stabilised after cold drawing is complete. At that stage the flow units will all be oriented in the direction of stress and the system is once again stable. The stress required to produce further strain increases becomes increasingly great, the curve once against sweeps upward and does so increasingly steeply. At this stage the applied force, having overcome the secondary cohesive forces of the original specimen earlier on, comes up against primary forces (molecular bonds, bond angles, hydrogen bonds, tightened entanglements). At the moment when instability would once again start, the high stress cannot be taken by a vulnerable spot in the same way as it could initially, because no reserve mechanism is left for this to happen, and rupture occurs. The point where the neck regains maximum stability can be determined by an extension of Considère's construction, as Vincent [1] has pointed out (Fig. 8.7).

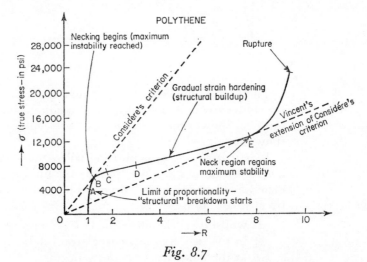

POLYTHENE

Fig. 8.7

The region between C and D is clearly a region of 'solid flow' during which maximum instability is maintained, whilst between D and E gradual strain hardening (structural build-up) takes place—the exact reverse of the gradual strain softening (structural breakdown) which had previously taken place between A and B. Point B lies in what corresponds to the pseudoplastic region and point E lies in what corresponds to the dilatant region of the generalised liquid flow curve.

We shall now leave the true stress curve and confine ourselves to conventional stress curves which are directly and therefore most readily determinable.

Not all thermoplastic polymers are able to stabilise the neck and whether the neck becomes thinner and breaks or stabilises itself depends not only on the polymer and other compounding ingredients, but also on the molecular weight of any one polymer itself. If the flow units are too small to form entanglements or can only form few and loose entanglements which easily slip apart in the 'solid flow' region, then rupture will occur at relatively low stresses because the stresses required for strain hardening cannot be supported. The amount of strain hardening decreases with the molecular weight of the flow units as can be seen qualitatively even from the conventional stress/strain curves in Fig. 8.8, which were obtained for polythenes of different melt flow indexes.

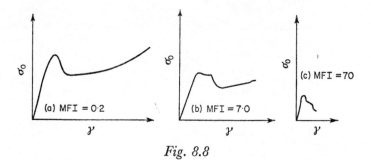

Fig. 8.8

The highest molecular weight material (lowest MFI) shows the full extent of cold drawing, medium molecular weight material shows much less and low molecular weight material shows none at all [2].

It is true that thermoplastics, when tested under arbitrary conditions do not necessarily show the full development of the generalised stress/strain curve. The curve may stop short because of rupture:

(*a*) whilst it is still within the Hookean proportionality limit,

(*b*) in the region of increasing strain softening,

(*c*) at the point where necking begins,

(*d*) during necking (as in Fig. 8.8*c*),

(*e*) during strain hardening but before the neck is fully stabilised.

In typical thermoset materials it is never possible to exceed the proportionality limit sufficiently to obtain necking because, owing to the density of cross linkages, viscous shear flow cannot get under way to such an extent as to dominate the stress response. But in lightly crosslinked thermosets (which are untypical plastics) and in many rubbers elastoviscous responses of the Voigt model type (see Chapter 7) may involve retarded dashpot flow. With filled phenolic and aminoplastic moulded specimens failure generally occurs within the proportionality limit and involves little deformation up to the rather high stresses necessary to produce rupture.

In thermoplastics, on the other hand, the stress/strain curve can take any form derived from the generalised curve; it should, however, be added that curves which fall progressively shorter of full development through premature rupture of the specimen are also progressively depressed towards the strain axis and in the extreme case the maximum itself will degenerate into an apparently insignificant knee. The actual curve obtained will depend on the severity of the experimental conditions (Fig. 8.9). In (*j*) no flow is observed, the response appears to be ideally elastic. In (*h*) some flow (shear) is

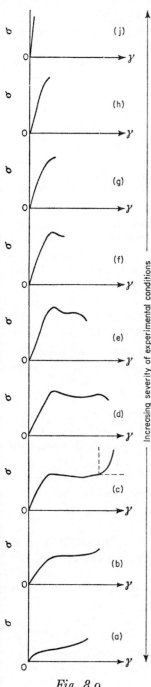

Fig. 8.9

indicated by derivation from linearity just before rupture, whilst in (*g*) dominance of flow deformation is just reached. Up to this point a material is said to fail in a *brittle* manner. From (*f*) to (*c*) permanent deformation involves progressively greater deformation and smaller stress at failure. Such materials are said to fail in a *tough* manner. The case of (*c*) deserves further comment. Drawn polymers such as nylon or polyester fibre, monofilament or film have had a previous stress history when they are being tested. What appears to be the starting point of the test (dotted coordinates) is not the true starting point at all. If the curvature of the stress/strain curve of, say, a drawn nylon monofilament shows a curvature 'the wrong way round' and fails in a brittle manner, this is entirely due to the fact that, after stress hardening, shear flow has ceased to be the dominant response and tension has reasserted its dominance. In (*b*) and finally in (*a*) the divide between tensile elastic and viscous flow has become rather smudged, the viscous forces tend to dominate and the deformation at failure may not be of a high order—the rubbery state has been reached.

Clearly, if a material fails in a brittle manner it fails in tension, and if it fails in a tough manner it fails in shear. The probability of brittle failure therefore increases with the severity of the test.

Tests can be made more severe in a number of ways:

1. The free volume can be decreased so that the viscous shear-deformational response to an applied stress becomes increasingly difficult. This is simply achieved by reducing the temperature, or by reducing the amount of plasticising additives (including low molecular weight fractions) in the compound.

2. The time available for the flow units to get the viscous deformational process under way and so to dissipate the imparted stress energy may be reduced so that the only response then open to the flow units is a tensile response. This is fairly readily done by increasing the

rate of test, in the extreme case up to impact conditions.

3. The applied stress can be chosen so that its tensile component is increased at the expense of the shear component. Indeed, if an exclusive triaxial tensile stress could be applied to a specimen (which is practically impossible), then the specimen would fail in a brittle manner whatever the state of the other variables. Triaxiality can be approached by notching the specimen increasingly sharply, as will be seen later.

4. The specimen may be given a previous stress history which has already traversed the region of flow instability so that the specimen has regained stability to a greater or lesser extent. Alternatively the specimen may contain stored energy in the form of frozen-in stresses which will make a contribution equivalent not only to the true total stress but which will also accelerate the rate of stress application when additional stresses are applied. This, clearly, is a function of processing conditions and is absent in properly annealed samples in which the stresses have been fully dissipated by allowing corresponding strains to develop.

5. The free volume could also conceivably be reduced by the application of very high pressures. It is likely that a tensile specimen tested whilst immersed in a hydraulic liquid under high pressure will fail in a brittle manner when the same specimen will fail in a tough manner at atmospheric pressure but at otherwise equal conditions.

6. The crosslinkage of polymer chains, whether achieved chemically (as in the case of unsaturated polyesters) or by irradiation, will produce a network. The crosslink density of this network will determine the degree to which solid flow under stress is restricted. The greater the flow restriction, the less stress can be dissipated and the greater will be the relative severity of a tensile test, so that brittle failure becomes increasingly probable.

7. The chemical nature of the polymer chains themselves can be modified either directly or by modification of the monomer, say by inclusion of comonomer to the monomer feed prior to polymerisation. This will promote tough rather than brittle failure. For example:

(i) The chlorosulphonation of polythene to 'Hypalon' increases free volume and destroys crystallinity. It also gives subsequent crosslinking possibilities with the result that a material is obtained which is a rubber at ordinary temperatures and which is 'tougher' than the original polythene.

(ii) Copolymerisation, e.g. of vinylidene chloride with butyl acrylate or of ethylene with vinyl acetate. This also destroys —at least partially—the crystallinity which is characteristic for poly(vinylidene chloride) and polythene respectively. The result in the case of the vinyl polymer is a corresponding increase in free volume and the conversion of an otherwise brittle, rather insoluble and thermally intractable polymer to a copolymer which is tough, tends to be elastomeric, is soluble in suitable solvents and—most important—which is thermally processable. This is known as 'internal plasticisation'. Many similar examples could be given. External plasticisation—as in flexible PVC—has the same effect, although it is physically rather than chemically induced.

Apart from the obvious effect of deliberate notching one may consider other situations where defects are most probably the root cause of brittle failure in otherwise tough materials:

1. The fact that polymers are, after all, molecular chains, implies that the inter-chain regions and especially chain ends constitute regions of weakness compared to the chains themselves. The regions of weakness at chain synapses may in fact be regarded as micro-defects on a molecular scale. These may be exaggerated in environments which are capable of further interfering with the cohesive forces at molecular interfaces and synapses, so that actual micro-cracks or crazes are developed during 'environmental stress cracking' (ESC). Polymers which consist predominantly of crystalline material will be less prone to ESC than those with substantial amounts of amorphous phase and this is borne out when the more highly crystalline linear (high density) polythenes are compared to the less crystalline (low density) polythenes. The latter are highly susceptible to surface active agents whilst the former are virtually immune to ESC in the presence of surface active agents. Again, polycarbonates (which are essentially amorphous) are highly susceptible to ESC by vapours of hydrocarbon solvents. It does not follow that *all* amorphous polymers are susceptible to ESC. PVC, for example, is remarkably resistant to ESC. But it would be fair to say that ESC would be likely to occur in amorphous rather than crystalline material if it occurs at all. If the length of flow units becomes so small, however, that each unit consists of a whole molecule the chain length of which is sufficiently small for that whole molecule to form a discrete unit of a multi-molecular crystal, then the synapses will cease to be points of attack for ESC, because there will then be little if any amorphous phase left. This is the case in *very* low molecular weight polythenes ('polythene waxes') and, in the extreme case, in paraffin wax. The increase in crystallinity—as the molecular weight reduces to that of paraffin wax—also reduces the probability of solid flow (viscous deformation) and the material will again fail in a

brittle rather than in a tough manner. Despite the fact that *intermediate* molecular weight polythenes have more amorphous phase than very low or very high molecular weight polythenes, the expected increase in toughness is more than counter-balanced by their relatively high susceptibility to ESC and consequent embrittlement through micro-defects.

2. The total energy required to produce brittle failure is really made up of two components: a crack *initiation* energy and a crack *propagation* energy. A material will always fail in the manner which corresponds to the lowest energy necessary to produce failure. If the energy necessary to produce brittle failure is higher than the energy necessary to produce tough failure (i.e. failure by permanent viscous flow deformation), then the specimen will fail in a tough manner, and vice versa. This provides a key for 'toughening' otherwise brittle materials. Crack initiation can be reduced by addition of plasticisers; crack propagation can be reduced by arresting the crack after it has formed. The former is achieved, for example, by addition of plasticiser to rigid PVC, the latter by the inclusion of a highly elastic disperse phase as for example in toughened polystyrene (Fig. 8.10). In Fig. 8.10*a* the force is providing

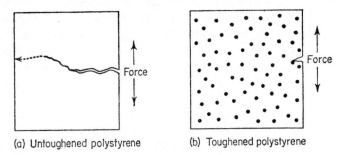

(a) Untoughened polystyrene   (b) Toughened polystyrene

Fig. 8.10

sufficient energy to initiate a crack and is propagating that crack across the specimen which is about to fail in a brittle manner. In Fig. 8.10*b* the same energy has merely initiated the crack but finely dispersed rubber particles are absorbing and dissipating the remaining energy, so that the specimen cannot fail in a brittle manner under the same conditions. The brittle strength has been raised above the tough strength and when failure does occur it will therefore be tough failure. The efficiency of toughening polystyrene in this way depends on:

(i) the probability of the propagating crack meeting a crack-arresting rubber particle; that is to say, it depends on the amount of rubber present and on the fineness of dispersion;

(ii) the intimacy of contact and the degree of adhesion of the interfaces between the phases so that the approaching crack is not deflected away from the energy-absorbing rubber particles by physical approach barriers such as voids;

(iii) the mass of each dispersed particle, since the energy absorbing capacity is necessarily a function of the mass of the particle. There is an optimum dispersion of such a fineness that the probability of a propagating crack being intercepted is high and the energy absorbing capacity for arresting the crack on interception is adequate. This optimum is experimentally determinable and it has been proved that 'over-fine' dispersion reduces the effective toughening obtainable from a constant weight proportion of disperse phase.

Unfortunately polystyrene loses its optical clarity by the inclusion of a disperse phase. The drilling of articles, of course, constitutes notching and one would very much like to toughen the impact-brittle drilled acrylic windscreen of a motorcycle which invariably cracks from the drill-hole when the machine is dropped. Unfortunately the inclusion of dispersed rubber is not feasible—although this might well achieve the object of overcoming shock brittleness in the notched windscreen—for the obvious reason that the excellent clarity which, in addition to its good weathering properties, led to the use of acrylic sheet for this application, would be lost.

3. The inclusion of fillers generally promotes embrittlement because it *also* constitutes notching of a kind. On the other hand:

(i) if fillers have compensating properties (such as energy absorption potential, as in dispersed rubbers, or the spreading of stresses over larger volumes, as in fibrous fillers),

(ii) or if they have specific chemical or physical interactions with the polymer (as is sometimes the case with special clays, silicas and carbons) which may outweigh the intrinsic notch embrittlement due to their inclusion, then these fillers may promote toughness in the nett result where embrittlement would otherwise have been the dominant effect.

It is now necessary to give a number of definitions which will enable us to make a more detailed examination of stress/strain curves:

Modulus: the slope of the tangent to the curve.

Yield point: that point on the conventional stress/strain curve where necking starts. This is the maximum of the curve where the modulus (tangent to the curve) is zero.

Yield strength: the stress per unit area at yield.

Brittle strength: the stress per unit area at break if the material breaks before it yields, i.e. if it fails in a brittle manner.

Ultimate strength: the stress per unit area at rupture (identical with the brittle strength in brittle materials, but different from the yield strength in tough materials).

Energy at yield: the area under the stress/strain curve ($\int \sigma\, d\gamma$) up to the yield point.

Energy at break: the total area under the stress/strain curve ($\int \sigma\, d\gamma$) from its origin up to rupture.

These are all parameters of the stress/strain curve. In addition there are the complementary parameters of *elongation at yield* and *elongation at break* which define themselves and which complete the list.

It is important to note that while the initial modulus is of fundamental significance (analogous to the 1st Newtonian viscosity in liquids), it is only a function of stiffness (rigidity, resistance to flow deformation) and cannot be a function of

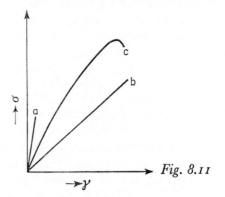

Fig. 8.11

strength. This is illustrated in Fig. 8.11. Material (*a*) has the highest initial modulus. It is the stiffest material. But whilst (*b*) is much less stiff than (*a*) it is obviously stronger because the stress at break is greater. Both (*a*) and (*b*) are brittle, but (*c*) which is tough is also stronger than both (*a*) and (*b*), intermediate in stiffness (initial modulus) between (*a*) and (*b*).

Obviously, the initial modulus (or for that matter, any modulus taken from the curve) has no connection with either the strength or the type of failure of the material. This is why static modulus determinations which are so useful in characterising thermal transitions and give so much structural information are of rather limited interest to the *design engineer* who is concerned with the stresses which a component can stand before it becomes unserviceable and who may perhaps also wish to know what deformation is involved should the material be tough.

A design engineer is only interested in *one* strength parameter. If the material fails in a brittle manner the brittle strength is the only available strength parameter anyway. But it is pointless to give the second (ultimate) strength in a material which is tough and has already yielded. Nobody will be interested in the ultimate strength of a PVC raincoat after it has already been pulled out of shape to such an extent that the owner cannot wear it any more. But the stress which it will stand before it yields obviously constitutes useful information. The only exception to this is for drawn fibre, film and monofilament. Here the previous stress history (including the yield point) is ignored and the ultimate strength is quoted—but the drawn material can obviously be regarded as a qualitatively different species from the undrawn material.

It is clear that, in order to assess the strength of a product one must test it to destruction, that is to say, to rupture in a brittle material and to yield in a tough material. This poses a problem: If one were to guarantee the impact strength of, say, a polyester/glass fibre crash helmet one would have to crush it under simulated contingency conditions and have nothing left to sell. Inevitably, one must test to simulated contingency conditions (also taking into account fatigue effects through repeated stress exposure), allow for a suitable safety factor and guarantee the stress endurance at that level. If one could be quite certain of quality consistency in production (a big if!) one would not need to test every specimen, but could content oneself with testing to destruction one out of an arbitrary number of mouldings.

The design engineer has an additional problem: Whilst data on strength parameters of materials prepared according to standard specifications abound, these may give an entirely misleading picture when they are applied to the actual component because:

1. The component will be different in size and shape;

2. It could conceivably have greater stiffness designed into it by suitable changes in its geometry;

3. It may have internal corners constituting notching to varying degrees of severity;

4. The level of residual stresses may be very different from those which a material test specimen moulded under arbitrary standard conditions may have. Furthermore, the distribution of residual stresses is almost always non-uniform;

5. The service temperature conditions may be different;

6. The stress levels which the component should withstand without failure may be very different;

7. If cylic stresses are involved their frequency may be very different;

8. The stress system (to some extent idealised in standard tests) may be different in either type or complexity.

As a result it is futile to base a product design on materials tests unless all the variables are fully covered and this is such a stupendous task that it could not be undertaken even for a single polymer type—remembering that each type comes in many grades of molecular weights, molecular weight distribution, fillers, plasticisers or other modifying additives—let alone for the whole spectrum of plastics. The best that can be done is to draw intelligent inferences from carefully chosen materials tests, select the possible materials, make the component and test it under actual service conditions in the hope that some correlation of ranking will exist between materials and product tests.

If this sounds like a counsel of despair, this is not altogether so—it is merely a warning against possible pitfalls and premature conclusions. In particular, no single test can characterise the strength of a plastic. Having accepted that, what *can* be done?

If the severity of a test can be changed by changing one of several experimental conditions while keeping all the others constant, then it is clear that identical stress/strain curves can be obtained in a variety of ways. This has some important implications: An impact test can only tell us qualitatively whether a material is shock brittle or not, and if the material *is* shock brittle then the impact test will also give the energy at break. Even that energy is rather inaccurate, however, since it makes no allowance for the kinetic energy imparted to the broken piece as it flies off, nor for the fact that the

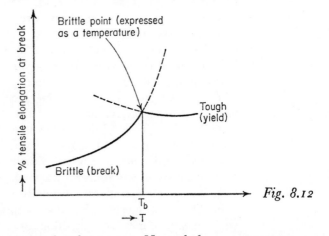

Fig. 8.12

clamping jaws absorb energy. Nevertheless, components are very commonly subjected to impact stresses during their functional life. In the short moment of impact it is obviously impossible to obtain a stress/strain curve. But if the same degree

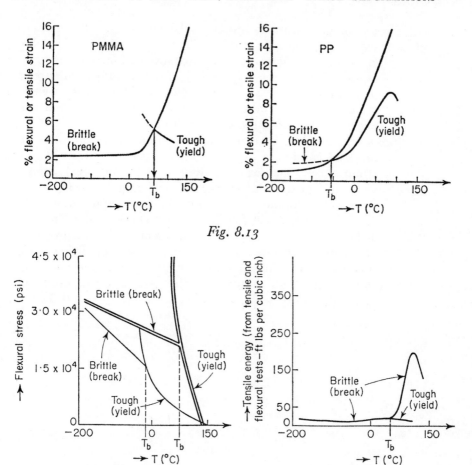

Fig. 8.13

Fig. 8.14

Fig. 8.15

of severity of test can be achieved by reducing the temperature, then the test need not be carried out under impact conditions at all; instead a curve obtained over a time interval which can be fully scanned for stress/strain relationships and which will give a result which is fully valid for predicting the behaviour of the material or component under impact conditions can be used. All that needs to be known additionally is what decrease in temperature is equivalent to a corresponding increase in the rate of stress application. In the same way notch embrittlement can be related to the embrittlement caused by an equivalent temperature reduction or by an increase in the rate of stress application. The *brittle point* can be expressed as the temperature at which the application of a given stress at a given rate (other variables also being constant) will produce a stress/strain curve which just terminates at the maximum, that is to say, at the tough/brittle transition, as in Fig. 8.9g. It can also be expressed in terms of stress application rate, degree of notching, etc., and thus acts as a watershed for the macroscopic type of failure. Indeed, the major problem in the study of the strength of plastics is to measure, understand and explain the trends of brittle strength whilst changing the temperature and other experimental variables one at a time.

A good way of exploring the failure characteristics of a material consists in plotting the strength parameters of stress/strain curves, such as elongation at yield and break, stress at yield and break, and energy at yield and break against the temperature or against the rate of stress application. Both yield and break will follow separate curves which cross over at the brittle point T_b (Fig. 8.12). At *that* point the probability of tough and brittle failure are exactly equal. Below T_b the probability of brittle failure will increase as the two curves (tough curve now dotted) diverge. Above T_b the probability of tough failure will increase analogously. Around T_b there

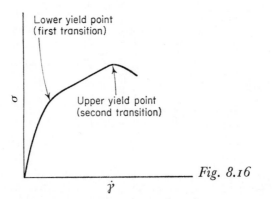

Fig. 8.16

will always be some specimens out of a suitably large number tested which fail in the untypical manner at much higher stresses and strains. The absolute trend of the curves is immaterial in this respect, the only thing that matters is their degree of divergence. Thus, while the elongation to *yield* in amorphous polymers (e.g. PMMA-polymethyl methacrylate) decreases along its observed course, that of crystalline polymers (e.g. PP-polypropylene) increases through most of its course (Fig. 8.13). But this does not affect the principal argument of tough/brittle transition.

The stress at break and yield also show a similar bifurcation which indicates the brittle point (Fig. 8.14) and so does the energy at yield and break (Fig. 8.15).

An exactly similar picture emerges when the *straining rate* is changed instead of the temperature. The tough/brittle transition will then be expressed in reciprocal seconds and the symbol $\dot{\gamma}_b$ for the transition will take the place of the symbol T_b.

Some materials have more than one apparent yield point (from the point of view of serviceability). In such a case (Fig. 8.16) the *lower* yield point is the relevant one for design considerations.

If the tensile straining rate $\dot{\gamma}$ in log form is plotted vs temperature, then a family of parallel straight lines is obtained each of which refers to a constant yield stress (Fig. 8.17) [4].

Results on rigid PVC in tension at seven different straining rates and six different temperatures produced curves from which the following numerical relationship was deduced for the material:

$$\sigma_{\text{yield}} = 14.4 - 0.119T + 1.12 \log \dot{\gamma} \; [4]$$

In laboratory tests specimens are usually subjected to 'simple' tests such as tension. In service the type of stress which puts the component at risk may be more complex.

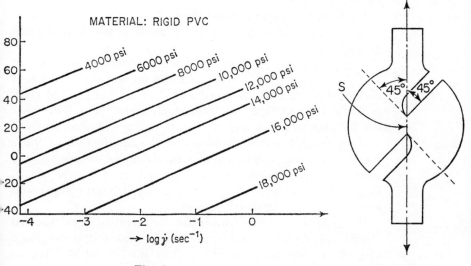

Fig. 8.17   Fig. 8.18 [5]

Obviously, one cannot test all materials for all possible stress systems, so one has to study the relationship between test results from 'simple' stresses with those obtained from complex stresses. The word 'simple' has been put into inverted commas because, as will be seen presently, considerable complexity may be hidden under the cloak of apparent simplicity.

The question which must be answered is: What type of stress systems are likely to promote tough and brittle failure respectively? The stress at any point in a loaded solid can be reduced to three-component tensile or compressive stresses. To simplify matters, let us assume that one of these three components is zero, so that we are left with two tensor components, σ_1 and σ_2. If a positive sign denotes tension and a negative sign compression, then it is clear that a two-component stress system can be devised in five ways:

1. Uniaxial compression $(-\sigma_1, 0)$.
2. Uniaxial tension $(+\sigma_1, 0)$.
3. Biaxial tension $(+\sigma_1 + \sigma)$.
4. Biaxial compression $(-\sigma_1 - \sigma)$.
5. Shear $(+\sigma/2, -\sigma/2)$.

We can ignore biaxial compression because it is a somewhat contrived rather than a naturally occurring stress system. The other four, one would imagine, can be simulated quite easily in the laboratory.

Uniaxial compression by compression of a specimen between two parallel plates;

Uniaxial tension in a tensile test;

Biaxial tension in a clamped membrane which is stressed at one point;

Shear by applying a uniaxial tensile stress at an angle of 45°[5] (Fig. 8.18). Note that a shear stress is equivalent to a combination of a tensile stress and a numerically equal compressive stress at right angles to each other. The B.S. shear test for plastics which specifies the punching out of a disc from a sheet specimen falls far short of the application of 'pure shear'. However, we know that in a uniaxial tensile test the specimen is simultaneously compressed in a lateral plane. Conversely, the skin of a specimen under uniaxial compression is simultaneously strained in tension.

Photoelastic patterns (strain lines in polarised light) showed clearly that the stress concentration was greatest in the region S (Fig. 8.18) and that the strain lines intersected the axis of locus S at an angle of 45°. Reasonably pure shear was thus ensured and the shear stress was, moreover, approximately constant over the cross sectional area at S.

However, we know that in a uniaxial tensile test the specimen is simultaneously compressed in a lateral plane. Conversely, the skin of a specimen under uniaxial compression is simultaneously strained in tension.

It is clear that:

either: shear is a composite system consisting of a uniaxial tensile and compressive component;

or: uniaxial tension (compression) is itself a complex system consisting of a shear and a biaxial tensile (compressive) component.

This can be graphically and mathematically expressed as follows [6]:

In Fig. 8.19a uniaxial tensile components were taken as the primary coordinates, whilst in 8.18b the coordinates have been rotated clockwise through 45° and the shear and biaxial tensile components are considered to be the primary coordinates. In the first case shear appears as a composite stress made up of uniaxial tension and compression whilst in the second uniaxial tension (compression) appears as a composite stress made up of shear and biaxial tension (compression). Either of these views is equally tenable: In the first case:

$$(+\sigma/2, -\sigma/2) = (\sigma_1, 0) + (0_1 - \sigma)$$

shear   uniaxial   uniaxial
tension   compression

and in the second

$$(+\sigma_1, 0) = (+\sigma/2, -\sigma/2) + (+\sigma/2, +\sigma/2)$$

uniaxial   shear   biaxial tension
tension

The actual *magnitude* of the stress (whether σ or $\sigma/2$) is unimportant because we are only concerned with a qualitative resolution at this stage.

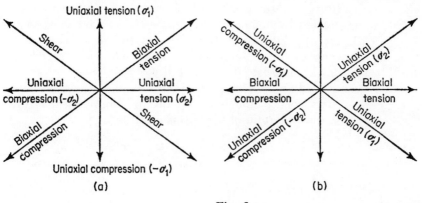

(a)   (b)

Fig. 8.19

Which of these two concepts has physical reality? The answer to this is provided when a tensile specimen of rigid PVC yields. Markings appear in the direction of shear and this is at an angle of *less* than 90° to the tensile stress axis. If these markings were at exactly right angles to the tensile stress axis, then the shear component would be zero. The angled pattern of parallel shear lines was first observed in mild steel under tension and is known as the 'Lüder bands'.

When a tensile specimen fails in a tough manner, it always fails in *shear*. Now, shear is a flow phenomenon—therefore a solid that yields is in reality a liquid of sorts. Moreover, once a shear force can cause solid flow (yield) the internal friction will generate heat. This in turn pushes the stress/strain curve in the same direction as if the test specimen temperature had been raised, with the result that failure must necessarily be tough.

If a tensile test is, however, carried out very rapidly (tensile impact conditions), that is to say, over a time interval which is shorter than the relaxation time of the plastic and shorter than the time necessary for the establishment of resonating waves, then the flow mechanism cannot get under way, no Lüder bands will appear and the specimen fails in a brittle manner with a fracture surface at right angles to the tensile stress axis.

If the shear component is great enough, then tough failure will always occur. The critical proportion of shear in the stress system which just produces tough failure varies from plastic to plastic and is of course also dependent on other variables.

Of course, one must consider the problem in all three dimensions. Notching will not only increase the straining rate, it will also affect the proportion of shear to tensile stress in a stress system: The stress at the base of a notch will tend to convert a uniaxial tensile stress $(\sigma_1, 0, 0)$ into a triaxial tensile stress $(\sigma_1, \sigma_2, \sigma_3)$ [7], which is due to the fact that the struts created by notching will oppose contraction at right angles to the applied stress. If we accept that it is *shear stress* which controls the yielding of the specimen, then we can see why notches increase the probability of brittleness:

(*a*) In a uniaxial stress system $(\sigma_1, 0, 0)$ the maximum tensile stress conceivable is σ and the maximum shear stress is $\sigma/2$, so that the ratio of the two is given by:

$$\frac{\text{tensile stress}}{\text{shear stress}} = \frac{\sigma}{\sigma/2} = 2$$

(*b*) On the other hand, in a *notched* specimen where the uniaxial tensile stress $(\sigma_1, 0, 0)$ has become converted to a triaxial tensile stress $(\sigma_1, \sigma_2, \sigma_3)$ where $\sigma_1 > \sigma_2 > \sigma_3$, the maximum conceivable tensile stress is σ_1 and the maximum shear stress is $\sigma_1 - \sigma_3/2$. In this case the ratio of the two is given by:

$$\frac{\text{tensile stress}}{\text{shear stress}} = \frac{2\sigma_1}{\sigma_1 - \sigma_3} \geqslant 2$$

In other words, notching has caused a reduction in the share of the shear stress within the total stress system, and since shear stress tends to produce tough failure the probability of brittle failure is increased. Obviously, therefore, notching should be avoided in plastics design, or at least reduced in severity by generous radiusing at the notch base.

Where a plastic is already brittle notching cannot cause further embrittlement although it will reduce the strength and the energy-to-break. But near the tough/brittle transition even a slight notch can swing the balance sufficiently to make a normally tough unnotched material fail in a brittle manner (e.g. rigid PVC at room temperature). If the brittle strength is *very much* greater than the tough strength, however, then even severe notching will not be sufficient to induce brittleness and all that happens is that the material will yield in the notch region. In so doing it will increase the notch radius and make the probability of brittle failure still more remote. This is the case, for instance, with low density polythene at room temperature even at deformation rates corresponding to free-fall impact.

One more complication in viscoelastic responses must be dealt with which has already been hinted at. One usually considers these responses to be associated with stress and strain rate tensors, but this presupposes that the tensors have sufficient time to resolve themselves into components, whatever the appropriate coordinate system may be [8]. If, however, this is not the case, then the relaxation processes (including stress distribution by stress-wave propagation) cannot get under way and the stress will remain anisotropic. Indeed, we

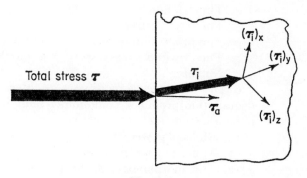

Fig. 8.20

may assign, for any time interval insufficient for resonating stress waves to become established along the length of a tensile test specimen, an anisotropic stress contribution τ_a to the

system, *in addition to* the *isotropic* stress contribution τ_i which latter is resolvable into components (Fig. 8.20):
Whilst

$$\tau = \tau_a + \tau_i = \tau_a + [(\tau_i)_x + (\tau_i)_y + (\tau_i)_z]$$

$$\frac{\tau_a}{\tau_i} = f(t)$$

where $f(t)$ tends to infinity as time tends to zero, and $f(t)$ tends to zero as time tends to infinity.

In order to clarify the relative proportion of anisotropic stress in the stress system it is necessary to evaluate τ_a/τ_i as a function of time. This cannot readily be done directly, but one can get *some* information by making use of the fact that brittle failure is associated with the absence of relaxation mechanism and also with the presence of notches which serve to increase the triaxiality of the stress. To put it another way: If notches make it sufficiently more difficult for the material to resolve a stress input into isotropic components, then the unresolved (anisotropic) portion becomes more important. Eventually the anisotropic portion predominates or becomes at any rate, sufficiently large to reduce the dissipation of stress significantly, so that, as a nett result, the probability of brittle failure is increased.

It was seen that the complexity of nonlinear viscoelastic response to large deforming stresses causes enormous difficulties in assessing plastics in a manner which is reliable and meaningful for design purposes. The following variables are involved:

1. Temperature
2. Strain and straining rate
3. Stress and stressing rate
4. Time (frequency)
5. The type of stress
6. The proportion of anisotropy in the total stress
7. The previous stress history (e.g. moulding stresses)
8. Notching
9 The presence of modifying additives (fillers, plasticisers, degradation products, comonomer built into the chain, cross-linking agents, etc.)
10. Molecular weight
11. Molecular weight distribution
12. The morphology of the polymer (crystallinity, size of crystallites, orientation)
13. Environmental stresses
14. Pressure

Faced with this formidable list one can nevertheless devise a system which gives useful design information without spending a lifetime on the evaluation of just one grade of one material. According to Turner [8] this should be based on the following experiments:

1. Two or three creep tests at different stresses at 20°C, followed by a study of recovery after creep;

2. An isochronous stress/strain curve* at 20°C at, say, 100 seconds on the same standard material as was used for the creep test.

One obtains a three-dimensional stress/strain/time response surface of a solid model which can be represented in planar fashion by means of contour lines with any two out of the three variables constituting the base coordinates. The isochronous stress strain curve represents the section through the solid model in a plane at right angles to the time axis.

3. A study of the effects of moulding stresses (and their removal by annealing) and of environmental effects such as humidity, ultraviolet light, etc., on the material by means of creep and isochronous stress/strain tests;

4. A more detailed study of the effects of *temperature* on
   (i) the isochronous stress/strain curve,
   (ii) creep curves at various stresses,
   (iii) the storage and loss modulus of the complex dynamic modulus at varying frequencies.

To these four main experiments the following information might be added:

5. Quotation of at least *some* shift factors caused by increasing the test severity in terms of their temperature equivalent, especially those for notching, time (frequency), type of stress, and stress history.

The next chapter will deal with the dynamic modulus and its significance in plastics evaluation and polymer research.

---

* An isochronous stress/strain curve can be obtained in the following manner:

(i) Applying an arbitrary constant strain rate and reading off the corresponding stress after a certain time interval (say, 100 seconds).

(ii) Repeating the process at various strain rates.

9 Dynamic testing—a tool for product evaluation and research

For convenience this is based largely on electrical volume elements (dipoles) subject to their appropriate dynamic stress field, but the treatment is fully applicable to any other volume element and its corresponding dynamic stress field (classical-mechanical, optical, magnetic, acoustic, etc.) and the analogies are pointed out. Complex moduli, storage and loss moduli (classical-mechanical, electrical, etc.). The Cole-Cole plot. Spectral deviation from the response predicted by the Debye Equations for the frequency dependence of storage and loss moduli and the causes of such deviation. The Cole-Cole plot as a means for determining distributions of relaxation times. The mechanism through which a loss modulus arises. Dielectric performance. Mechanical performance. Dynamic glass transition and its dependence on temperature/time (frequency) and additives, implying time/temperature superposition and free volume concepts. Instruments for dynamic mechanical testing: the torsion pendulum and the rolling ball spectrometer. Summary on the uses of dynamic spectra (especially electrical and mechanical loss spectra) for structural elucidation and product performance.

A DYNAMIC test is any test which subjects a volume element to a stress which varies with time. But recently the term has acquired a more specific meaning and one now regards a dynamic test as one in which a stress is applied which varies periodically in magnitude. The most common type of periodic variation is sinusoidal and this also lends itself readily to mathematical treatment. In Chapter 7 reference has been made to relaxation and retardation times, as well as to complex, storage and loss moduli. Both the model theory and the phenomenological theory of linear viscoelastic behaviour can be applied to dynamic stresses and the Debye equations for the storage and loss moduli were given. The importance of cyclic (sinusoidal) stresses and strains has been increasingly recognised in the mechanical field as it has become apparent that they play a decisive part in the failure of components and assemblies through fatigue. They have been used for some time in the optical field, not only in the visible but also in the infra-red, ultra-violet and X-ray regions for the purpose of research and especially for chemical and structural analysis. They are the mechanism of acoustic transmission. They have played a dominant part in electrical conduction and insulation.

Indeed, the fundamental mathematics has been worked out from electrical theory and the conclusions reached were found to be fully applicable to sinusoidal stresses and strains in general, irrespective of the size of the volume elements or the nature of the deformation which may be involved.

Just as an elastic modulus and loss modulus (viscosity) can be determined on liquids, so the storage and loss modulus can be determined in solids. The two moduli represent the components of the complex dynamic modulus. The electrical elasticity modulus ε' is known as the 'permittivity' or 'dielectric constant', whilst the electrical loss modulus ε'' is known as the 'dielectric loss'. Both are functions of the applied field frequency and of the dipole relaxation time of the material under the test conditions. The two moduli are given by the Debye equations [1, 2]:

Equations 9.1 and 9.2

$$\varepsilon' = \varepsilon_\infty + \frac{\varepsilon_s - \varepsilon_\infty}{1 + \omega^2 \tau^2}, \quad \varepsilon'' = \frac{(\varepsilon_s - \varepsilon_\infty)\omega\tau}{1 + \omega^2 \tau^2}$$

where ε_∞ = the dielectric constant at infinity frequency
ε_s = the static dielectric constant (zero frequency)
ω = frequency in radians
τ = relaxation time

Equations 9.1 and 9.2 can be written:

Equation 9.3

$$\frac{\varepsilon' - \varepsilon_\infty}{\varepsilon_s - \varepsilon_\infty} = \frac{1}{1 + \omega^2 \tau^2}$$

and

Equation 9.4

$$\frac{\varepsilon''}{\varepsilon_s - \varepsilon_\infty} = \frac{\omega\tau}{1 + \omega^2 \tau^2}$$

If we plot the left-hand sides of Equations 9.3 and 9.4 vs log ω we obtain the following symmetrical curves (Fig. 9.1):

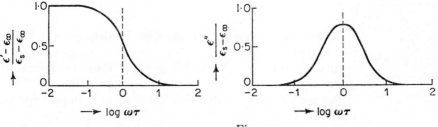

Fig. 9.1

The loss peak for ε'' is reached when $\log \omega\tau = 0$, or when $\omega\tau = 1$, that is to say, when the frequency is equal to the reciprocal of the relaxation time.

The largest value for the dielectric constant ε' and loss ε'' can be obtained directly from Equations 9.1 and 9.2 by putting $\omega\tau = 1$, when

Equation 9.5
$$\varepsilon'_{max} = \frac{\varepsilon_s + \varepsilon_\infty}{2}$$

Equation 9.6
$$\varepsilon''_{max} = \frac{\varepsilon_s - \varepsilon_\infty}{2}$$

A neat method of checking whether the Debye equations are rigorously obeyed is due to Cole and Cole. It consists of plotting ε'' vs ε' at constant temperature with varying frequency:

Equation 9.7
$$\varepsilon' - \frac{\varepsilon_s - \varepsilon_\infty}{2}^2 + (\varepsilon'')^2 = \frac{\varepsilon_s - \varepsilon_\infty}{2}^2$$

Thus, by plotting ε'' vs ε', a semicircle should be obtained with radius $(\varepsilon_s - \varepsilon_\infty)/2$. Its centre must be on the ε' axis (abscissa) at a distance $(\varepsilon_s + \varepsilon_\infty)/2$ from the origin. The points of intersection of the semicircle with the abscissa are those for which ε' is numerically equal to ε_s and ε_∞. Assuming that $\varepsilon_s = 10$, $\varepsilon_\infty = 2$ and $\tau = 10^{-10}$ sec, the following graph would be obtained (Fig. 9.2):

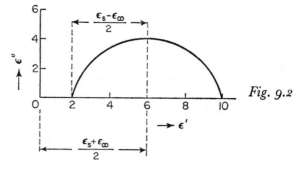

Fig. 9.2

Note that ε'' and ε' are completely defined if ε_s and ε_∞ are known and that the Cole-Cole plot is independent of relaxation time which does not enter into the relationship directly.

For the total area under the $\varepsilon''/\ln \omega$ curve a simple expression can be derived:

Using the identity:

$$\int_0^\infty \frac{d\omega}{1 + \omega^2\tau^2} = \frac{\pi}{2\tau}$$

we obtain from Equation 9.2:

Equation 9.8
$$\int_0^\infty \varepsilon'' \frac{d\omega}{\omega} = \frac{\pi}{2}(\varepsilon_s - \varepsilon_\infty)$$

This enables us to calculate $(\varepsilon_s - \varepsilon_\infty)$ when ε'' is known over the entire frequency range.

Experimental results are often expressed in the form of the ratio $\varepsilon''/\varepsilon'$ as a function of frequency. This ratio, also known as $\tan \delta$ (see later) follows from the Debye equations:

Equation 9.9
$$\tan \delta = \frac{\varepsilon''}{\varepsilon'} = \frac{(\varepsilon_s - \varepsilon_\infty)\dfrac{\omega\tau}{1 + \omega^2\tau^2}}{\varepsilon_\infty + \dfrac{\varepsilon_s - \varepsilon_\infty}{1 + \omega^2\tau^2}} = \frac{(\varepsilon_s - \varepsilon_\infty)\omega\tau}{\varepsilon_s + \varepsilon_\infty\omega^2\tau^2}$$

On differentiating Equation 9.9 with respect to ω and setting $d\tan\delta/d\omega$ to zero, we see that the maximum value of $\tan \delta$ is obtained at a critical frequency given by:

Equation 9.10
$$\omega_c = \frac{1}{\tau}\sqrt{\frac{\varepsilon_s}{\varepsilon_\infty}}$$

and is equal to:

Equation 9.11
$$(\tan \delta)_{max} = \frac{\varepsilon_s - \varepsilon_\infty}{2\varepsilon_s\varepsilon_\infty}$$

The corresponding values ε_c' and ε_c'' at which $\tan \delta$ is at a maximum are obtained from the Debye equations substituting ω_c (as per Equation 9.10) for ω:

Equation 9.12
$$*\varepsilon_c' = \frac{2\varepsilon_s\varepsilon_\infty}{\varepsilon_s + \varepsilon_\infty}$$

Equation 9.13
$$*\varepsilon_c'' = \frac{\varepsilon_s - \varepsilon_\infty}{\varepsilon_s + \varepsilon_\infty}\sqrt{\varepsilon_s\varepsilon_\infty}$$

In practice experimental curves for ε' and ε'' vs log ω deviate considerably from the values predicted by the Debye equations. The ε' curve is flatter and extends over a wider frequency range and the ε'' curve is broader and has a lower maximum than that predicted by Equation 9.6. Generally, however, the curves still remain symmetrical (Fig. 9.3).

* Note that ε_c' and ε_c'' are different from ε'_{max} and ε''_{max}.

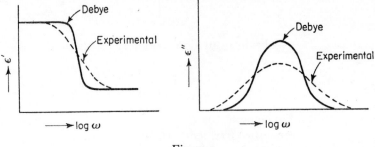

Fig. 9.3

As a consequence, the Cole-Cole plot also deviates from the semi-circular. Cole and Cole showed that the experimentally obtained ε'' vs ε' plots still tend to be circular, but that the centre of the circle lies under the abscissa. The reason for this is that the idealised model is too simple. It is just not true that for each molecule the magnitude of the interacting forces, the magnitude of the applied stress and the temperature are constant throughout the bulk of the test specimen. Each dipole has its own relaxation time and is also affected by neighbouring dipoles. This is true for relatively simple molecules, but even more so for complex ones. In measurements of moduli one therefore obtains an average value, a most probable value, around which other values will cluster.

If $G(\tau)$ is the distribution function of relaxation times, then $G(\tau)\,d\tau$ represents the fraction of volume elements associated, at a given instant, with relaxation times which range from τ to $(\tau + d\tau)$. $G(\tau)$ is 'normalised' such that the total area under the distribution curve is equal to unity:

Equation 9.14

$$\int_0^\infty G(\tau)\,d\tau = 1$$

The modified Debye equations are:

Equation 9.15

$$\varepsilon' = \varepsilon_\infty + (\varepsilon_s - \varepsilon_\infty)\int_0^\infty \frac{G(\tau)\,d\tau}{1 + \omega^2\tau^2}$$

and

Equation 9.16

$$\varepsilon'' = (\varepsilon_s - \varepsilon_\infty)\int_0^\infty \frac{\omega\tau G(\tau)\,d\tau}{1 + \omega^2\tau^2}$$

Using Equation 9.16 it can be shown that Equation 9.8 for the total area under the $\varepsilon''/\ln \omega$ curve is also valid for a material which has a *distribution* of relaxation times:

Equation 9.17

$$\int_{-\infty}^\infty \varepsilon''\,d\ln \omega = (\varepsilon_s - \varepsilon_\infty)\frac{\pi}{2}$$

Cole and Cole showed that for a distribution of relaxation times the equation for the complex modulus ε^*, namely:

Equation 9.18
$$\varepsilon^* = \varepsilon_\infty + \frac{\varepsilon_s - \varepsilon_\infty}{1 + i\omega\tau}$$

must be modified to:

Equation 9.19
$$\varepsilon^* = \varepsilon_\infty + \frac{\varepsilon_s - \varepsilon_\infty}{1 + (i\omega\tau_0)^{1-h}}$$

where τ_0 and h are constants.

If there is only *one* relaxation time and not a distribution, then $h = 0$ and Equation 9.19 reduces to Equation 9.18. The parameter h is a function of the subtending angle ϕ of the circle sector in the Cole-Cole plot (Fig. 9.4), namely:

$$\phi = \pi - 2\alpha$$

and

$$\alpha = \frac{\pi h}{2}$$

so that

$$\phi = \pi(1 - h)$$

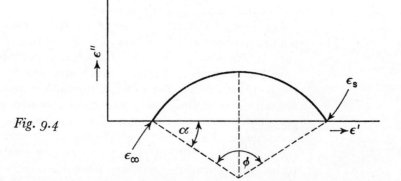

Fig. 9.4

Unfortunately for structural analysis the distribution of relaxation times in plastics is rather scattered and h is large (the limiting value is unity). This means that the Cole-Cole circle shows above the ε' horizon only to the extent of a shallow segment which is easily missed. Moreover, because there may be *a number of species* of dipoles, the distribution of their relaxation times may overlap and the resulting smudging of the Cole-Cole segments tends to make the clear identification of individual segments exceedingly problematical (Fig. 9.5). The continuous line is the experimental observation whilst the dotted lines represent hypothetical distributions of relaxa-

183

tion times of which only the arrowed one stands out, and that one is obviously *not* coincident with a true loss maximum. Misinterpretation is therefore all too easy.

What is unfortunate for the structural analyst is, however, a great advantage for plastics generally. If h is very large indeed, say, close to its limiting value of unity, then the Cole-Cole segment degenerates into a tangential point and the exponential term in the denominator of Equation 9.19 becomes

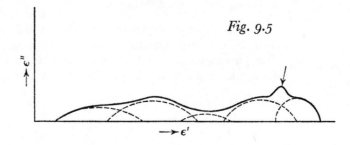

Fig. 9.5

unity. This means that the imaginary (loss) portion of the complex modulus (ε'') disappears and $\varepsilon^* \equiv \varepsilon_s$. h is close to unity in plastics and therefore they are not very lossy electrically, in fact they are excellent insulators.

However, excellence is only a relative term. Looking at plastics in isolation, it soon becomes obvious that these materials vary greatly amongst themselves. Clearly, polar polymers will have loss peaks at frequencies which are close to the mean of the reciprocal relaxation times of the dipoles involved. There are not many commercially important polymers which are *not* polar by nature—the list is virtually complete with polyolefines, polystyrene, PTFE and polyvinylidene chloride. Amongst the more highly polar polymers are the polyamides and PVC, whilst the polyesters, acrylics, phenolics, aminoplastics and epoxides have weaker dipoles. Oxidative degradation in polyolefines can increase the vanishingly small loss factor of polythene by two or three decades and thus affords a means of assessing the amount of antioxidant present in the compound. Polar impurities (especially catalyst residues in stereospecific polyolefines) greatly increase the magnitude of dielectric losses. This is particularly serious in long distance cable insulation and it is probably true to say that only the availability of highly sophisticated ultra-low-loss polythenes of power factor 0·00002 to 0·0001 have made the land-sea cable from Britain to Australia possible. Even small losses are serious because the low thermal conductivity of plastics makes it difficult for the heat which is generated in the process to be dissipated. As the heat builds up, so the relaxation times decrease, the material may become more lossy and eventually thermal breakdown may occur. But even losses can be put to good use: PVC can be high-frequency welded, phenolics and aminoplastics can be preheated by applying a dielectric field of suitable frequency for a few seconds. Preheating reduces moulding cycles, makes for better flow, lower residual stresses, reduced mould wear and more uniform mouldings of thermosetting moulding materials such as PF and the aminoplastics.

It is now appropriate to examine briefly the mechanism which causes the loss modulus to arise.

Polarity exists in certain chemical bonds, especially carbon to halogen and carbon to nitrogen, giving rise to large dipole moments. Dipole moments arising from carbon to hydroxyl and carbon to oxygen bonds are also clearly detectable. The polarity arising from carbon to hydrocarbon substituent groups (alkyl, aryl, cycloaromatic) are, however, very small. The higher the polarity, the greater the susceptibility of a dipole to align itself to conform to an applied electric field. This field will then further distort the dipole. Indeed, it may even induce the formation of dipoles where dipoles do not exist in the ground state.

If the electric field is static, then all the energy of polarisation is stored in the distorted dipole which acts like a spring and reverts to its rest position when the field is removed. But if the field alternates in polarity, then the dipole will tend to move with the field and in so doing it will do work against the frictional resistance (internal viscosity) of its environment. Heat is generated and energy is lost. The 'brake' on the oscillating dipoles will become more and more noticeable as the field frequency is gradually increased and is a direct function of the phase difference between the field and the lag in the response. There will exist a certain frequency when this phase lag is at a maximum, whereupon, on further increasing the frequency, the dipoles will become less and less sensitive to field changes because they are less and less able to respond within the decreasing time interval available between changes of polarity. The frequency at which the lag (loss) is greatest is around the inverse relaxation time of the dipole. The dielectric constant (storage) is highest at zero frequency, that is to say, when the field is static, and least at very high frequencies, i.e. respectively before and after the lossy region has been traversed. Maxwell has observed that the dielectric constant at very high frequencies in nonpolar liquids is very nearly equal to the square of the optical refractive index:

$$\varepsilon_\infty' \cong n^2$$

Electrically the storage modulus (dielectric constant) ε' is defined as the ratio of the capacitance C of the dielectric and the capacitance of the empty space C_0 of the same dimensions; the loss modulus is a measure of the resistance current going through the dielectric and is given by

$$\varepsilon'' = \frac{I}{\omega R C_0} = 36\pi 10^{11}\frac{K}{\omega}$$

where R = resistance of the dielectric in ohms

K = specific conductance in ohm^{-1} cm^{-2}

C_0 = vacuum capacitance in Farads (ohm^{-1} sec)

In a simple capacitor the capacitative current I_c and the resistive current I_R are given by:

$$I_c = i\omega\varepsilon'C_0E \quad \text{and} \quad I_R = \omega\varepsilon''C_0E$$

where $E = E_0 \cos \omega t$ and E_0 = the amplitude of the alternating voltage E. The relationship between total current I, capacitative and resistive currents are given in Fig. 9.6.

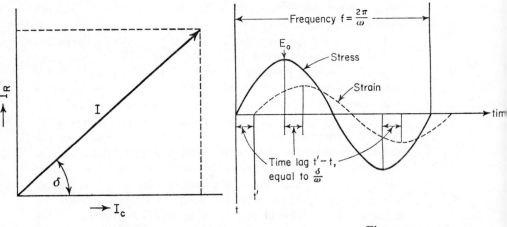

Fig. 9.6   Fig. 9.7

It is clear that I_R/I_c is numerically equal to $\varepsilon''/\varepsilon'$ and when plotted as in Fig. 9.6 the ratio can be regarded as the tangent of an angle δ. The significance of the angle δ becomes apparent when the stress and the strain response are plotted together as a function of time (frequency). It is clear that the stress leads the strain by a constant phase difference which can be expressed in terms of radians (Fig. 9.7) and the phase angle δ. This remains constant since the strain follows the stress at the same frequency.

The study of the electrorheology of plastics is a fascinating subject but since there is no space for a detailed treatment which alone would do it justice the reader is strongly urged to consult the following excellent and very simply presented references: (1) Hoffman: The mechanical and electrical properties of polymers, *I.R.E.* (*Component Parts*), June 1957, **CP-4**, 2, 42–69; (2) Mikhailov and Sazhin: Macromolecular dielectrics, *Russian Chem. Reviews* (trans. into English), July 1960, **29**, 7, 410–419; (3) Mathes in Baer, Engg. Design for Plastics, Ch. VII, Reinhold 1964.

Briefly, the following features should be noted:

1. The dielectric constant is a function of the electronic polarisability α_e and density ρ in nonpolar materials and is given by the Lorenz–Lorentz equation through the optical refractive index n:

$$\frac{\varepsilon' - 1}{\varepsilon' + 2} \cong \frac{n^2 - 1}{n^2 + 2} = \frac{4\pi N}{3M} \alpha_e \rho$$

where N = the Avogadro Number, $6 \cdot 023 \times 10^{23}$, and
M = molecular weight.

It follows that the electronic polarisability is inversely proportional to the density. This means that in crystalline materials where the density changes abruptly at the melting point T_m, and in amorphous materials where the density changes at different rates with temperature around T_g, the transitions are indicated by dielectric constant measurements as a function of temperature, and that these transitions are frequency-independent. It is a useful exercise to consider these transitions with the free volume concept well in mind.

2. The very small loss ε'' of nonpolar polymers is still unexplained. Extensive milling on hot rolls say, of polythene in the presence of air causes substantial losses because of the appearance of carbonyl groups, but it is difficult to ascribe the small basic loss to this or similar forces or to impurities alone.

3. In polar materials there is much greater scope for electrical investigation of dipole structures from which chemical and morphological details may be inferred. Here we have a permanent dipole moment the orientation of which is measured by the *orientational* polarisability α_μ:

$$\alpha_\mu = \frac{\mu^2}{3kT}$$

(where k is the Boltzmann constant).

This is additional to the *electronic* polarisation α_e which is *always* present. The contribution of α_μ can raise the total polarisation to very high levels. For highly polar materials we have:

$$\varepsilon_s - \varepsilon_\infty = \frac{32\pi N\rho}{3M}\frac{\mu^2}{3kT}$$

This means that:

(a) If μ is large, $\varepsilon_s - \varepsilon_\infty$ may be very large.

(b) $(\varepsilon_s - \varepsilon_\infty)$ varies inversely with temperature. Hence the value of ε_s tends to decrease markedly as the temperature increases.

Why does the orientational polarisability (and hence ε_s) depend so strongly on temperature in polars, and why is the temperature dependence of the inverse type? The thermal agitation interferes with the orientating effect of the field—at high temperatures the dipoles already move vigorously because they are thermally energised and the relatively gentle field influence cannot therefore make as great a contribution to ground-state-departure as at lower temperatures where the thermal agitation is much less pronounced.

4. The orientational polarisability depends strongly on the ability of the dipoles to turn in a field. In a liquid this can readily manifest itself, but less so in an amorphous solid and even less so in a solid in which the dipoles are firmly locked in a crystallite. This means that, although ε_s tends to reduce with increasing temperature, the unlocking effect when crystals melt at T_m will reverse the downward trend of $\mathrm{d}\varepsilon_s/\mathrm{d}T$ and overwhelm it. The same argument also applies qualitatively to the glass transition in amorphous polymers.

Polar side groups, if present, can make a distinct contribution to the overall polarisability which thus does not have to be restricted to the main chain. There can therefore exist additional glass transitions which refer specifically to dipoles other than main-chain dipoles. These can be readily picked out over the temperature spectrum of dielectric loss.

5. The dielectric relaxation times which characterise the loss maxima can be of the order of 10^{-10} sec in the liquid state, but at low temperatures they can be very long and will therefore only show up at low frequencies. In order to obtain the overall picture one must therefore traverse many decades of frequencies. This can fairly readily be done electrically and makes dielectric measurements a powerful research tool.

6. The activation energy E for dipole orientation processes can be obtained from a plot of $\log \tau$ vs $1/T$ which, in conformity with the Arrhenius equation gives a slope of $2\cdot303E/R$. The τ value used is that obtained from a Cole-Cole plot or an ε'' vs $\log \omega$ plot for each of a series of temperatures. The

measured activation energy in polars usually lies between 5 and 50 kcal/mole. A significant point is that the activation energy for some polar polymer crystals is not very different from that found for dipole orientation in systems which consist of randomly distributed (amorphous) chain molecules. This proves that the *whole* chain need not move to permit the required degree of polar mobility. The higher the activation energy, the greater is the change of the relaxation time with temperature.

7. As plasticiser is added, so the relaxation time becomes shorter. This is entirely expected from the free volume concept: the dipoles are freer and the loss maximum occurs at higher frequencies. This shifting in the loss maximum by means of plasticisation is sometimes very useful in driving the loss maximum from certain frequency ranges in which one wants to use an insulating polymer but is prevented from doing so by excessive dielectric loss. The same 'plasticising' effect can also be achieved by judicious copolymerisation. But the first step in such 'tailoring' is an exact knowledge of the lossy regions (see p. 184).

The dielectric moduli have been very fully investigated for a number of polar polymers, including poly(ethylene terephthalate) [3] and poly methyl methacrylate. On the basis of these measurements solid models and contour maps of the dielectric constant/temperature/frequency and dielectric loss/temperature/frequency relationships have been constructed.

A number of attempts have been made to relate polymer structure with the dynamic electrical or dynamic mechanical properties [4, 5, 6].

Deutsch, Hoff and Reddish, for instance, point out that the crystallites of poly(ethylene terephthalate) can have units arranged in a symmetrical trans-form and in an asymmetrical cis-form:

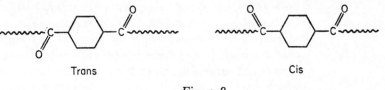

Trans   Cis

Fig. 9.8

The polyester has a dielectric loss peak around 80°C which is also near its mechanical glass transition at which temperature crystallisation begins to occur on reducing the temperature. It is therefore reasonable to suppose that the dielectric loss is associated with a transition in the amorphous phase which

189

becomes impossible in the crystalline phase. That the packing of chains is similar in the two phases is indicated by the fact that the density difference is only about 10 per cent. The benzene rings are probably in parallel planes locally and the oxygens are coplanar with the benzene rings. In the crystalline phase the large dipole moment of the cis form is frozen (so is the trans-form, but in the trans-form the dipole moments are in opposite directions and therefore cancel out). In the amorphous phase, on the other hand, some distortion in response to an electric field is likely to occur and at the absorption peak the cis-trans transition which is associated with the freedom of movement due to T_g occurs frequently enough for the repeating unit to have a measurable average dipole moment. The relaxation time of this dipole moment must be related to the average time of the transition process.

Optical and electrical loss spectra are fairly familiar, but it has been demonstrated for a number of polymers that *mechanical* loss peaks are also characteristic. On the whole, mechanical spectra are of rather limited use in structural research. Firstly, the available frequency range is incomparably smaller than for optical and electrical spectra; and secondly the applied stress is rather indiscriminate in the response it excites. A gross mechanical stress will not only affect mechanical flow units, but also such electrical or magnetic dipoles as may be present, as well as the chemical bond geometries. Since the latter three involve much more subtle changes their effect will be swamped by the mechanical macro-effect. The subtle changes of the nonmechanical responses can only be isolated by applying correspondingly subtle and highly specific modes of excitation. Mechanical and electrical excitations are qualitatively the same thing, but only in the sense that a tornado is qualitatively the same as a gentle zephyr. A zephyr will rustle the leaves of a tree, a tornado will undoubtedly do the same, but this will not be very remarkable considering the violent vibrations and swayings of twigs, branches and even the trunk itself. On the whole, each mechanical loss region (in polar compounds at least) also corresponds to an electrical loss region and both tend to occur in the same temperature and frequency regions. This is not surprising, since both are associated with flow units which become immobile as the temperature drops (or as the frequency increases). Whether an electrical loss peak occurs in a mechanical loss region depends, of course, on whether the flow unit contains dipoles or not.

Polymers containing water may be regarded as two-phase systems. The water may act as a plasticiser (as in polyamides), but some water may also be present in polymers which are intrinsically hydrophobic, as in polystyrene and PVC, especially if hygroscopic additives are present. It will therefore contribute to the comparatively high dielectric loss in emulsion polymers when these are compared with mass, solution or suspension polymers. In the case of polyamides the increase in conductance is so great that one must look for a second effect. This is provided if one assumes that the polyamides also act as polyelectrolytes.

In order to assess the electrical suitability of a plasticiser one must carry out the measurements at an isoviscous point, that is to say, at the various temperatures at which the flexibility of all compositions is the same. One will then prefer the plasticiser which has the fewest number of dipoles per unit volume of plasticised compound whilst yet having an acceptable cold flex temperature.

We must now leave the border region of interplay of mechanical and electrical loss spectra and consider the mechanical loss spectra more fully. We have already seen that mechanical spectra can contribute to an understanding of polymer structures and morphology but it has been pointed out that the scope is limited because of the limitation to conveniently realisable frequency ranges. However, those frequency ranges that *are* available are also those which are important for the mechanical and acoustic performance of plastics engineering components.

If a simple polymer is cyclically stressed at a given temperature and frequency it will respond in one of two ways:

Either it will behave as an elastic rubber with a modulus of about 10^7 cgs units,

Or it will behave as a glass with a modulus of about 10^{10} cgs units, with little deformation occurring.

If the time between segmental movements is short compared to the time of the stress period, then the material will comply readily. This is the case for the rubbery condition. The reverse will apply to the glassy condition.

Obviously, if the segmental periodic time decreases with increasing temperature, a transition will occur beyond which the material will behave as a rubber and this transition will occur within the narrow limits around T_g. But if the periodic time of stress is decreased (frequency increased), then the transition will occur at a higher temperature. Therefore, as Karas point out [7], it is possible to make a rubbery material behave as a glass by striking it very quickly: An ordinary impact blow takes 1–2 milliseconds and under such a blow the material will behave as a rubber; but if the same rubber is struck at 1 microsecond it may shatter like a glass. Again we

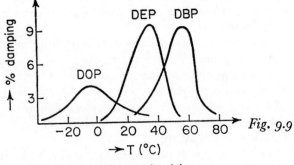

Fig. 9.9

PVC with different plasticisers
(after Neilson)

have time/temperature superposition and if we operate in
the linear viscoelastic region then this will apply even more
rigorously than in the nonlinear region of viscoelastic behaviour
which was the subject matter of Chapter 8.

If additives are introduced, several possible effects may occur,
depending on the nature of the additive. The addition of
varying amounts of plasticiser to PVC will alter the position
of the transition region. Alteration in the type of plasticiser
will change the form and position of the transition (Figs. 9.9
to 9.12).

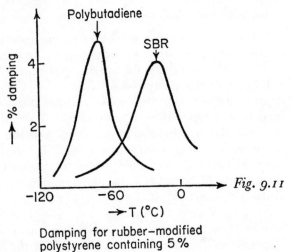

Fig. 9.11

Damping for rubber–modified
polystyrene containing 5%
added rubber

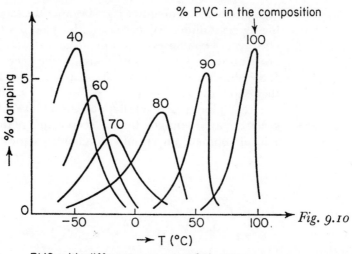

PVC with different amounts of the same
plasticiser (DOP)(after Wolf)

With rubber modified PSt the addition of rubber may produce a new molecular species with its own damping peak in
the same way as do solvated complexes in simple molecules.
Whether it does or not will depend on the type and quantity
of rubber used.

A change in molecular weight will not generally alter the
dynamic properties, except at quite low molecular weights.
It should be clear why this must be so. But a change in the
regularity or in the configuration of side groups on a chain
will result in a change of the spectrum. Atactic polymers differ

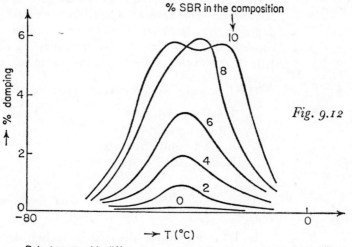

Polystyrene with different amounts of the same added rubber (SBR)

greatly from stereoregular ones because the latter are generally highly crystalline. Where crystallinity occurs the material has effectively two phases—amorphous and crystalline—and separate transition regions will therefore be seen in the spectra. If the phases of multiphase systems are completely incompatible, then separate transitions will be found for each phase. Where the two phases are totally and mutually soluble, then the spectrum approximates that of a copolymer of the two materials. Partial compatibility results in a complex spectrum.

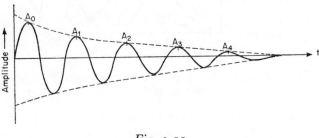

Fig. 9.13

The work on dynamic mechanical spectra involves frequencies below 10^4 cycles per second, mostly less than 2,000 cycles per second. At such frequencies simple apparatus is available. Although forced vibrations give perfectly acceptable results, the simplest apparatus relies on free oscillation techniques. A sample is slightly strained and on releasing the stress the period of oscillation and the decay in amplitude of stress is measured. The rate at which the oscillations will die away depends on 'damping', i.e. the loss due to internal friction (internal viscosity, irreversible deformation, absorbed energy). It is therefore the precise analogue of the dielectric loss ε''.

The following instruments are used:

(*a*) Torsion pendulum
(*b*) Vibrating reed
(*c*) Ball rebound apparatus
(*d*) Rolling ball loss spectrometer.

Various versions of all of them are commercially available and allow measurements to be made over a wide temperature range.

(*a*) The operations involved in evaluating dynamic mechanical storage and loss moduli with a *torsion pendulum* (Fig. 9.14, p. 196) are as follows:

1. Determination of the resonance frequency ω and logarithmic decrement of the damped vibration Λ.

The amplitude of vibration at any time t is given by the equation

$$A(t) = A_0\, e^{-\omega \Lambda t} \cos (2\pi\omega t)$$

This function takes the graphical form as shown in Fig. 9.13. ω is determined from the graph by measuring the wavelength. Λ is simply

$$\ln \frac{A_0}{A_1} = \ln \frac{A_1}{A_2} \ldots = \ln \frac{A_n}{A_n + 1}$$

If the number of each peak is plotted vs $\ln A$ a straight line is obtained the slope of which is Λ.

2. The shear modulus G' (dynes cm^{-2}) is given by

$$G' = \frac{4\pi^2\omega^2 Il}{g}$$

where $I =$ torsional moment of inertia of the pendulum,

$\quad\quad g =$ a geometrical constant dependent on the cross sectional area of the specimen. For a rectangular cross section $g = bd^3C$, where

$\quad\quad\quad\quad b =$ breadth,

$\quad\quad\quad\quad d =$ thickness,

$\quad\quad\quad\quad C =$ a dimensionless function of the ratio d/b which is curvilinear and is usually supplied with the instrument.

$\quad\quad l =$ free length of specimen between clamps.

The damping factor ($\tan \delta$) is simply given by

$$\tan \delta = \frac{\Lambda}{\pi}; \quad\quad \therefore G'' = G' \frac{\Lambda}{\pi} = \frac{4\pi\omega^2 \Lambda Il}{g}$$

3. Determination of the modulus and damping at constant frequency over a wide temperature range, at constant temperature and at several frequencies. The frequency can be varied by varying the inertial masses on the arms of the torsion bar or torsion disc.

4. Calibration of the torsion bar or torsion disc for the moment of inertia.

For the theory of calculation of the mechanical damping of polymers the reader is referred to Illers and Jenckel [8]. A torsion pendulum is very suitable for low frequency work.

(b) A *vibrating reed* is capable of giving results at much higher frequencies (audio frequency range) but requires sophisticated and therefore expensive electronics for recording. Highly elaborate apparatus is available which can be programmed to give a continuous record over a preset rate of temperature change using forced vibrations and measuring the energy loss

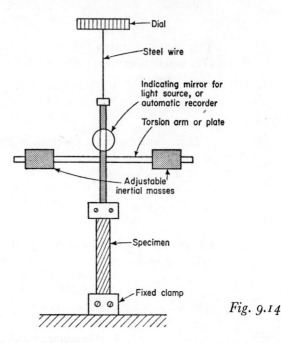

Fig. 9.14

which is represented by the power input necessary to keep the amplitude of the vibration constant.

(c) A *ball rebound tester* [9] is a simple device and can therefore be correspondingly cheap. It is especially useful for elastomeric polymers and film. A ball is released from a fixed height h_0 and strikes the specimen with a total energy E_t. The rebound height h is a direct function of the reflected energy E_r and the absorbed energy E_a is therefore

$$E_a = E_t - E_r$$

The absorbed and reflected energy are related in the same way as the loss and storage components of the complex modulus:

$$G^* = G' - iG'' = \text{const.} \times (E_a + E_r)$$

The ratio E_a/E_r therefore represents the loss (damping) factor tan δ.

The disadvantage of a rebound tester lies mainly in the technical difficulty of obtaining more than one rebound per drop due to a slightly off-vertical ball rebound, so that the ball may not drop back onto the specimen or may hit the walls of the glass cover. This limits the measurement to tan δ over a temperature range and precludes the measurement of

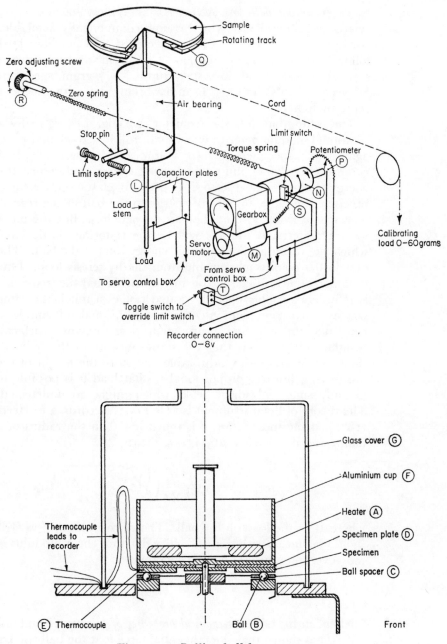

Fig. 9.15. Rolling ball loss spectrometer.

logarithmic decrements from which the real part of the complex modulus is determined.

(*d*) A *rolling ball loss spectrometer* is a most ingenious yet relatively simple device which is now commercially available. An instrument of this type was developed by Cheetham [10] following work by Flom in 1960. Cheetham's instrument can scan the mechanical loss spectrum of a 5-gram specimen between −120°C and 300°C in a working day with excellent reproducibility.

The rolling ball loss spectrometer (see Fig. 9.15) is really a rotational modulometer in which balls roll in a track and press against the specimen surface at the same time. The ball compresses polymer units which store energy and return the energy to the ball as it leaves. A perfectly elastic body returns precisely the same amount of energy to the ball at the rear as it exacts in front, so that there is no nett loss. But the more lossy a material is the less will be the restoring elastic force which the material returns to the energy total of the ball. The nett loss is readily measurable from the hysteresis loop. One of the great advantages of this instrument is that the frequency is infinitely variable over the stated range in much the same way as $\dot{\gamma}$ in a cone/plate viscometer and this makes it simple to construct the temperature/frequency/loss response surfaces. Another advantage is that the static pressure of the balls on the specimen is readily adjustable and since the height of the loss peak is linearly proportional to that load it is possible to magnify small subsidary peaks which might go undetected. The theory of the instrument is as follows: According to Hertz (1881), if the indentation d is much less than the radius of a ball r, and if Poisson's ratio is 0·5, then:

$$G' = \frac{3}{16} \frac{W}{\sqrt{r}} \left(\frac{1}{d}\right)^{3/2}$$

where W is the load on the ball. Therefore a plot of d vs $W^{3/2}$ should be linear, with slope G'. The dynamic shear modulus is, moreover, one-third of Young's modulus E:

$$G' = \tfrac{1}{3}E$$

Nielsen, in his book *Mechanical properties of polymers* (Reinhold, 1962), has shown that the tangential force F at the ball track is

$$F = K \frac{G''}{G'} \left\{\frac{1}{G'r^2}\right\}^{1/3} W^{4/3} \quad \text{where} \quad \frac{G''}{G'} = \tan \delta$$

K is a characteristic value for each polymer, which depends on Poisson's ratio. Poisson's ratio for rubbers is about 0·5; to this corresponds a value for K of 0·107, so that

$$F = 0 \cdot 107 \tan \delta \left(\frac{1}{G' r^2} \right)^{1/3} W^{4/3}$$

At constant load both $W^{4/3}$ and r are constant and F is therefore directly proportional to $\tan \delta$.

We have seen how dynamic spectra can be used to elucidate polymer structure, to evaluate the suitability of polymers for dielectric insulation and to supply mechanical data which are related to the performance of plastics under cyclic stresses.

Since the vibrations, rotations and other motions of chemical bonds are also highly frequency-selective it is not surprising that they show up as distinct spectrum. Since many types of motions involving chemical bonds are possible, even in relatively simple molecules, infra-red spectra are very complex, but the peaks have been accurately charted and they are fully indicative of the chemical nature of materials. Infra-red spectroscopy is a big subject and is dealt with in a number of standard reference works. No detailed account of optical spectra can be given here.

Electron spin and nuclear magnetic resonance spectra have a more recent history but their discussion lies also beyond the scope of this book.

Acoustic spectra can supply useful information on sound damping and one may expect more work to be done in this field in the future.

It is clear that dynamic spectra are of absorbing interest in every sense of the word.

The reader is strongly urged to study the contents of *Kolloid Zeitschrift*, December 1953, **134**, 2/3, in particular the papers by Würstlin, by Schmieder and Wolf, and by Staverman.

Appendix I: Damped inertial elasticity*

A mathematical treatment showing how a dynamic stress/strain situation arises as a special case of a general stress situation due to the relative magnitude of elastic and viscous responses coupled to the magnitude of inertial mass.

This appendix shows the mathematical connection between static and dynamic stress/strain situations.

Model:

The solution of the equation for damped inertial elasticity (see Equation 7.28, Chapter 7) is obtained as follows:

Equation A.1
$$\sigma = m\frac{d^2\gamma}{dt^2} + \eta\frac{d\gamma}{dt} + G\gamma$$

The general solution of this equation is the sum of (the particular integral plus the complementary function CF) where CF refers to the case of main interest and applies to zero stress, so that:

Equation A.2
$$m\frac{d^2\gamma}{dt^2} + \eta\frac{d\gamma}{dt} + G\gamma = 0$$

Take as a trial solution of (A.2):

Equation A.2a
$$\gamma = e^{\alpha t}$$

where α is obtained thus: Substitute for γ in (A.2):

$$m\alpha^2 e^{\alpha t} + \eta\alpha e^{\alpha t} + G e^{\alpha t} = 0 \quad \text{or} \quad e^{\alpha t}(m\alpha^2 + \eta\alpha + G) = 0$$

Now, $e^{\alpha t}$ can *never* be zero, therefore $m\alpha^2 + \eta\alpha + G = 0$, whence the two roots of α are:

Equation A.3
$$_1\alpha_2 = \frac{-\eta \pm \sqrt{\eta^2 - 4Gm}}{2m}$$

Now let:

Equations A.4

$$-\frac{\eta}{2m} = a \quad \text{If } \eta^2 > 4Gm \text{ the roots will be real.}$$

$$\frac{\sqrt{4Gm - \eta^2}}{2m} = b \quad \text{If } \eta^2 < 4Gm \text{ the roots will be complex and conjugate, i.e.:}$$

---

* For the mathematical treatment given below the author wishes to thank his colleagues of the Mathematics Department of the Borough Polytechnic, especially Mr Toms, who has since taken up another appointment at the Northern Polytechnic, London.

Equation A.5 $$\alpha_2 = a \pm ib$$

Case 1

(This is the most interesting case because, as will be seen, this leads to sinusoidal oscillations.)

Substitute for α (as per (A.5) in Equation A.2a):

Equation A.6 $$\gamma_1 = e^{(a+ib)t} \quad \text{and} \quad \gamma_2 = e^{(a-ib)t}$$

Note: If $_1\gamma_2$ are non-identical solutions of the differential equations, then:

(i) $A\gamma_1$ and $B\gamma_2$ are also solutions of the differential equations, and

(ii) The sum of any two solutions is again a solution of the differential equations, i.e.

Equation A.7 $$A\gamma_1 + B\gamma_2 = \gamma$$

Let us modify Equation A.6:

$$\gamma_1 = e^{at} \cdot e^{ibt}, \qquad \gamma_2 = e^{at} \cdot e^{-ibt}$$

Substituting in (A.7) we get for our most general solution:

$$\begin{aligned}
\gamma &= A\, e^{at} e^{ibt} + B\, e^{at} e^{-ibt} \\
&= e^{at}(A\, e^{ibt} + B\, e^{-ibt}) \\
&= e^{at}A(\cos bt + i \sin bt) + B(\cos bt - i \sin bt) \\
&= e^{at}[(A + B)\cos bt + i(A - B)\sin bt]
\end{aligned}$$

Let $A + B = E$ and $i(A - B) = F$, then:

Equation A.8 $$\gamma = e^{at}(E \cos bt + F \sin bt)$$

To solve (A.8) for any particular value of t (i.e. to obtain $\gamma(t)$), we fit our boundary conditions:

(i) $\gamma = \gamma_0$ when $t = 0$

Equation A.9 (ii) $\dfrac{d\gamma}{dt} = \Gamma$

(a known value which *could* be—but is *not necessarily*—zero when $t = 0$).

Setting $t = 0$ in (A.8):

Equation A.10 $$\gamma = \gamma_0 = e^0(E \cos 0 + F \sin 0) \quad \text{or} \quad \gamma_0 = E$$

Differentiating (A.8) we get:

$$\frac{d\gamma}{dt} = a\, e^{at}(E \cos bt + F \sin bt) + e^{at}(-bE \sin bt + bF \cos bt)$$

and when $t = 0$, this becomes:

$$\frac{\mathrm{d}\gamma}{\mathrm{d}t} = \Gamma = a\,\mathrm{e}^0(E\cos 0 + F\sin 0) + \mathrm{e}^0(-bE\sin 0 + bF\cos 0)$$

or

Equation A.11 $$\Gamma = aE + bF$$

We now solve the simultaneous Equations A.10 and A.11 to get

Equations A.12
$$\begin{cases} E = \gamma_0 \\ F = \dfrac{\Gamma - \alpha\gamma_0}{b} \end{cases}$$

Now that the constants E and F are known we substitute back into (A.8) and obtain the general solution for γ as a function of time:

Equation A.13
$$\gamma(t) = \mathrm{e}^{at}\left(\gamma_0\cos bt + \frac{\Gamma - a\gamma_0}{b}\sin bt\right)$$

where everything is known, remembering the definitions of a and b (see Equation A.4).

Case 2

Returning to Equation A.4: The condition for $\alpha_1 \equiv \alpha_2$ is that the root in Equation A.3 should disappear, that is to say, that:

$$\eta^2 = 4Gm \quad\text{or}\quad \frac{\eta^2}{G} = 4m$$

and since $\left(\dfrac{\eta}{G} = \tau\right)$ that $\tau = \dfrac{4m}{\eta} = -\dfrac{2}{a}$

In this case b becomes zero and Equation A.13 modifies as follows: The second term in the brackets on the right-hand side is multiplied and divided by t to become

$$t(\Gamma - a\gamma_0)\frac{\sin bt}{bt}$$

where $\dfrac{\sin bt}{bt} \to 1$ when $b \to 0$

Thus (A.13) becomes:

Equation A.14 $$\gamma(t) = \mathrm{e}^{at}[\gamma_0(1 - at) + \Gamma t]$$

and if Γ is also zero, then $\gamma(t) = \gamma_0\,\mathrm{e}^{at}(1 - at)$.

Equation A.14 can also be derived in the following way: Starting from (A.3) it is obvious that, for $\alpha_1 \equiv \alpha_2$, $\alpha_1 = -(\eta/2m)$. Then, formulating the most general solution corresponding to Equation A.7 (*Case 1*):

$$\gamma = A\,e^{\alpha_1 t} + B\,e^{\alpha_2 t}$$

which can be shown to be also

Equation A.15
$$\gamma = A\,e^{\alpha_1 t} + Bt\,e^{\alpha_1 t} = e^{\alpha_1 t}(A + Bt)$$

when $\alpha_1 \equiv \alpha_2$.

Boundary conditions: for $t = 0$ and $d\gamma/dt = \Gamma$, as before, and $\gamma = \gamma_0$.

Differentiating (A.15):

$$\frac{d\gamma}{dt} = \Gamma = \alpha_1\,e^{\alpha_1 t}(A + Bt) + e^{\alpha_1 t}.B$$

Setting $t = 0$ in (A.15) it follows that $\gamma = \gamma_0 = A$ and substitution in the *differentiated* Equation A.15 gives:

$$B = \Gamma - \alpha_1\gamma_0$$

Resubstituting for A and B in (A.15) then gives:

Equation A.16
$$\gamma(t) = e^{\alpha_1 t}[\gamma_0(1 - \alpha_1 t) + \Gamma t]$$

which is identical with (A.14), since, for the case $\alpha_1 \equiv \alpha_2$, α_1 is also identical with the a of Equations A.4 and A.14.

Case 3

If $\eta^2 > 4Gm$, we have distinct real roots:

$$\alpha_2 = \frac{-\pm\sqrt{\eta^2 - 4Gm}}{2m} = a \pm b$$

The general solution is:

Equation A.17
$$\gamma = A\,e^{\alpha_1 t} + B\,e^{\alpha_2 t}$$

At $t = 0$,

$$\gamma = \gamma_0 = A + B \quad \text{and} \quad \frac{d\gamma}{dt} = \Gamma$$

Differentiating (A.17),

$$\frac{d\gamma}{dt} = \Gamma = \alpha_1 A\,e^{\alpha_1 t} + \alpha_2 B\,e^{\alpha_2 t}$$

Hence

$$\begin{cases} A = \dfrac{\Gamma - \alpha_2\gamma_0}{\alpha_1 - \alpha_2} \\[2mm] B = \gamma_0 - A = \gamma_0 - \dfrac{\Gamma - \alpha_2\gamma_0}{\alpha_1 - \alpha_2} \end{cases}$$

Resubstituting for A and B in (A.17):

$$\gamma = \frac{\Gamma - \alpha_2 \gamma_0}{\alpha_1 - \alpha_2}\, e^{\alpha_1 t} + \left(\gamma_0 - \frac{\Gamma - \alpha_2 \gamma_0}{\alpha_1 - \alpha_2}\right) e^{\alpha_2 t}$$

or

$$\gamma(t) = \frac{\Gamma - \alpha_2 \gamma_0}{\alpha_1 - \alpha_2}\,(e^{\alpha_1 t} - e^{\alpha_2 t}) + \gamma_0\, e^{\alpha_2 t}$$

By inspection of (A.3:)

$$\left\{\begin{matrix} \alpha_1 = a + b \\ \alpha_2 = a - b \end{matrix}\right\}$$

$\alpha_1 - \alpha_2 = 2b$, so that we get:

Equation A.18
$$\gamma(t) = \frac{\Gamma - (a - b)\gamma_0}{2b}\,(e^{(a+b)t} - e^{(a-b)t}) + \gamma_0\, e^{(a-b)t}$$

(*Check:* At $t = 0$, Equation A.18 becomes: $\gamma(t) = \gamma_0$—correct.)

So far we have assumed, in trying to solve Equation A.1, that $\sigma = 0$, that is to say, that the stress decays *instantaneously* to zero at $t = 0$, i.e. on release of the static stress. But this is patently incorrect because, if the strain varies sinusoidally with time, so will the stress, although the stress will be out of phase with the strain owing to retardation. We know two basic facts about our model:

(i) That it is 'lossy', that is to say, that the complex modulus G^* can be split into a storage modulus involving recoverable deformation (the 'real' part of the complex modulus), and a loss modulus involving irrecoverable deformation, or flow (the 'imaginary' part of the complex modulus. This means that out of the three cases that may be considered we are principally interested in *Case 1*, where α_1 and α_2 are complex conjugate, i.e. $\alpha_2 = a \pm ib$ (Equation A.5).

(ii) The most general expression for the sinusoidal time/strain dependence of the stress is: $\sigma = e^{-kt} \sin \omega t$, where k is a function of m, η and G, namely the *stress decay function*, and where ω is the frequency of oscillations round the equilibrium position.

We therefore confine ourselves to *Case 1* and do *not* peruse other possible solutions ($\alpha_1 \equiv \alpha_2$, *Case 2*, or $_1\alpha_2$ both real distinct roots, *Case 3*), although students are advised, for the sake of manipulative expertise, to work out the solutions for the other two cases also. Note however, that only in *Case 1* (in which complex conjugate roots are involved) do sine functions enter. The solution applies to dynamic problems. If the roots are real, however, we are dealing with a static situation.

The general solution must include the particular integral which has been neglected up to now: $\gamma(t) =$ solution of complementary function (CF) + solution of particular integral (PI). Having defined the general stress function as

$$\sigma(t) = \mathrm{e}^{-kt} \sin \omega t$$

we can now write:

Equation A.19
$$m\frac{\mathrm{d}^2\gamma}{\mathrm{d}t^2} + \eta\frac{\mathrm{d}\gamma}{\mathrm{d}t} + G\gamma = \mathrm{e}^{-kt} \sin \omega t$$

where $k > 0$.

Using the symbol D for the differential operator $\mathrm{d}/\mathrm{d}t$ we can rewrite (A.19):

$$mD^2\gamma + \eta D\gamma + G\gamma = \mathrm{e}^{-kt} \sin \omega t$$

whence

$$\gamma = \frac{1}{mD^2 + \eta D + G}\, \mathrm{e}^{-kt} \sin \omega t$$

By the *shift theorem*, this can be shown to be:

$$\gamma = \mathrm{e}^{-kt}\,\frac{1}{m(D-k)^2 + \eta(D-k) + G}\, \sin \omega t$$

Using the theorem that:

$$\frac{1}{f(D^2)} \sin \omega t = \frac{1}{f(-\omega^2)} \sin \omega t$$

$$\gamma = \mathrm{e}^{-kt}\,\frac{1}{-\omega^2 m + (\eta - 2km)D + mk^2 - \eta k + G}\, \sin \omega t$$

or

Equation A.20
$$\gamma = \mathrm{e}^{-kt}\frac{1}{pD + q}\, \sin \omega t$$

where:

$$\begin{cases} p = mk^2 + G - \eta k - \omega^2 m \\ q = \eta - 2km \end{cases}$$

Multiplying top and bottom of the right-hand side of the last equation for γ by $(pD - q)$, we get

$$\gamma = \mathrm{e}^{-kt}\frac{pD - q}{p^2 D^2 - q^2}\, \sin \omega t$$

and using the previous theorem again:

$$\gamma = \mathrm{e}^{-kt}\frac{pD - q}{-\omega^2 p^2 - q^2}\, \sin \omega t$$

We have now eliminated the operator D from the *denominator* and only have it in the numerator, so that we can now use it directly to differentiate:

Equation A.21
$$\gamma = e^{-kt} \frac{1}{-\omega^2 p^2 - q^2} (-q \sin \omega t + p\omega \cos \omega t)$$

This, then, is the solution of the particular integral.

We now combine this solution with that given by (A.8) for the *CF*:

Equation A.22
$$\gamma = e^{at}(E \cos bt + F \sin bt)$$
$$- \frac{e^{-kt}}{\omega^2 p^2 + q^2} (p\omega \cos \omega t - q \sin \omega t)$$

We must re-evaluate E and F, using the usual boundary conditions, namely:

$$\gamma = \gamma_0 \quad \text{and} \quad \frac{d\gamma}{dt} = \Gamma$$

when $t = 0$.

Setting $t = 0$ in (A.22) we get:

$$\gamma_0 = E - \frac{1}{\omega^2 p^2 + q^2} p\omega$$

from which

$$E = \gamma_0 + \frac{p\omega}{p^2\omega^2 + q^2}$$

Differentiating (A.22):

$$\frac{d\gamma}{dt} = \Gamma = a\, e^{at}(E \cos bt + F \sin bt) +$$
$$+ e^{at}(-bE \sin bt + bF \cos bt) +$$
$$+ \frac{k\, e^{-kt}}{p^2\omega^2 + q^2} (p\omega \cos \omega t - q \sin \omega t) -$$
$$- \frac{e^{-kt}}{p^2\omega^2 + q^2} (-p\omega^2 \sin \omega t - q\omega \cos \omega t)$$

Substituting for E we get:

$$F = \frac{1}{b} \left\{ \Gamma - a\gamma_0 - \frac{\omega}{p^2\omega^2 + q^2} [p(a + k) + q] \right\}$$

Reintroducing E and F into Equation A.22 we obtain:

Equation A.23
$$\gamma = e^{at} \left\langle \left(\gamma_0 + \frac{p\omega}{p^2\omega^2 + q^2} \right) \cos bt + \right.$$
$$\left. + \frac{1}{b} \left\{ \Gamma - a\gamma_0 - \frac{\omega}{p^2\omega^2 + q^2} [p(a + k) + q] \right\} \sin bt \right\rangle -$$

(*contd.*)

$$-\frac{e^{-kt}}{p^2\omega^2 + q^2}(p\omega \cos \omega t - q \sin \omega t)$$

where p and q have been defined as per (A.20), a and b as per (A.4), and where all these are functions of m, η and G. k is also $f(m, \eta, G)$ and is the assumed stress decay function.

This leaves us with just one final problem—the evaluation of k. This is easy as a consequence of the experimental fact that the log of the ratio of successive amplitudes in a constant-frequency decay curve is constant:

$$\log \frac{h_1}{h_2} = \log \frac{h_2}{h_3} = \text{etc.} = \log \frac{h_n}{h_{n+1}} = \text{const.} = \Lambda$$

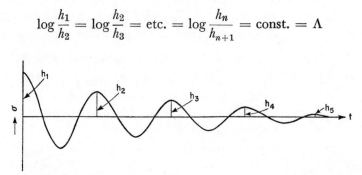

Λ is known as the mechanical loss factor or *damping factor*. Now, if $\sigma(t) = e^{-kt} \sin \omega t$, then

$$\sigma\left(t + \frac{2\pi}{\omega}\right) = e^{-k(t+2\pi/\omega)} \sin \omega t,$$

$$\log \sigma(t) = -kt + \log_n \sin \omega t,$$

$$\log_n \sigma\left(t + \frac{2\pi}{\omega}\right) = -k\left(t + \frac{2\pi}{\omega}\right) + \log_n \sin \omega t$$

Subtracting the last two equations for:

$$\log_n \sigma(t) \quad \text{and} \quad \log_n \sigma\left(t + \frac{2\pi}{\omega}\right)$$

$$\log_n \sigma(t) - \log_n \sigma\left(t + \frac{2\pi}{\omega}\right) = \log_n \frac{\sigma(t)}{\sigma\left(t + \frac{2\pi}{\omega}\right)}$$

$$= \log_n \frac{h_1}{h_2} = \text{const.} = \Lambda$$

and

Equation A.24

$$\Lambda = -\frac{2\pi k}{\omega} \quad \text{whence} \quad k = -\frac{\Lambda\omega}{2\pi}$$

We can now finally rewrite Equation A.23:

Equation A.25

$$\gamma(t) = e^{at}[U \cos bt + V \sin bt] - W e^{-kt}[p\omega \cos \omega t - q \sin \omega t]$$

where

$$U = \gamma_0 + \frac{p\omega}{p^2\omega^2 + q^2}, \qquad W = \frac{e^{-kt}}{p^2\omega^2 + q^2}$$

$$V = \frac{1}{b}\left\{\Gamma - a\gamma_0 - \frac{\omega}{p^2\omega^2 + q^2}[p(a + k) + q]\right\}.$$

U, V, W, a, b, p, q and k are all defined functions of the experimentally determinable or initially given parameters, namely:

$$m, G, \eta, \gamma_0, \omega, \Gamma \text{ and } \Lambda$$

Note the similarity of the fraction $p\omega/p^2\omega^2 + q^2$ with the characteristic terms of the Debye equations (Chapter 7) which are involved whenever a dynamic situation of the stress/strain relationship exists.

Appendix II: Solution of the differential equations for stress relaxation and strain retardation

1. *STRESS RELAXATION* (Maxwell Element):

$$\frac{1}{G}\frac{\mathrm{d}\sigma}{\mathrm{d}t} + \frac{\sigma}{\eta} = 0$$

$$\frac{\mathrm{d}\sigma}{\mathrm{d}t} = -\frac{G\sigma}{\eta}$$

$$\frac{\mathrm{d}\sigma}{\sigma} = -\frac{G}{\eta}\,\mathrm{d}t$$

$$\int\frac{\mathrm{d}\sigma}{\sigma} = -\int\frac{G}{\eta}\,\mathrm{d}t$$

$$\ln\sigma = -\frac{G}{\eta}t + \text{const.}$$

When $t = 0$, $\sigma = \sigma_0$

$\therefore$ const $= \ln\sigma_0$

$\therefore \quad \ln\sigma = \ln\sigma_0 - \dfrac{G}{\eta}t$

$\therefore \quad \sigma = \sigma_0\,\mathrm{e}^{-\frac{G}{\eta}t}$, where $\dfrac{\eta}{G} = \tau$, the *relaxation* time,

$\therefore \quad \sigma = \sigma_0\,\mathrm{e}^{-t/\tau}$

q.e.d.

2. *STRAIN RETARDATION* (Voigt-Kelvin Element):

$$\eta\frac{\mathrm{d}\gamma}{\mathrm{d}t} + G\gamma = \sigma.$$

in any equation of the form $\dfrac{\mathrm{d}y}{\mathrm{d}x} + Py = Q$, where both P and

Q are functions of x, one multiplies with an 'integration factor'

$\mathrm{e}^{\int P\mathrm{d}x}$, to obtain:

$$\mathrm{e}^{\int P\mathrm{d}x}\left(\frac{\mathrm{d}y}{\mathrm{d}x} + Py\right) = \mathrm{e}^{\int P\mathrm{d}x}\,Q.$$

on integration with respect to x:

$$\int e^{\int P dx}\left(\frac{dy}{dx} + Py\right) dx = \int e^{\int P dx} Q \, dx.$$

The left hand side is then: $y \, e^{\int P dx}$ (a standard integral). In the case of the equation which we wish to solve $P = \dfrac{G}{\eta}; \; Q = \dfrac{\sigma}{\eta};$ $y = \gamma; \; x = t;$ and $\dfrac{G}{\eta}$ and $\dfrac{\sigma}{\eta}$ are both functions of t.

Therefore, our integration factor is $e^{\int \frac{G}{\eta} dt}$, which is equal to $e^{\frac{G}{\eta}t}$. Multiplying with this integration factor after dividing by η, our equation becomes:

$$e^{\frac{G}{\eta}t}\left(\frac{d\gamma}{dt} + \frac{G}{\eta}\gamma\right) = e^{\frac{G}{\eta}t}\frac{\sigma}{\eta}$$

integration then gives:

$$\gamma \, e^{\frac{G}{\eta}t} = \frac{\sigma \eta}{\eta G} e^{\frac{G}{\eta}t} + \text{const.}$$

when $t = 0$, $\gamma = 0$,

$$\therefore \text{const} = -\frac{\sigma}{G}$$

$$\therefore \; \gamma \, e^{\frac{G}{\eta}t} = \frac{\sigma}{G} e^{\frac{G}{\eta}t} - \frac{\sigma}{G}$$

$$\therefore \quad \gamma = \frac{\sigma}{G} - \frac{\sigma}{G} e^{-\frac{G}{\eta}t}$$

$$\therefore \quad \gamma = \frac{\sigma}{G}\left(1 - e^{-\frac{G}{\eta}t}\right), \text{ where } \frac{\eta}{G} = \tau, \text{ the } \textit{retardation} \text{ time,}$$

$$\therefore \quad \gamma = \frac{\sigma}{G}\left(1 - e^{-\frac{t}{\tau}}\right)$$

q.e.d.

References

Chapter 1

1. A. K. Van der Vegt and P. P. A. Smit, 'Crystallisation Phenomena in Flowing Polymers', paper read at the Conference on Advances in Polymer Science and Technology, London, Sept. 1966.
2. J. Galt and B. Maxwell, *Mod. Plastics*, Dec. 1964., 115, etc.
3. H. Green and R. Weltmann, *Ind. Eng. Chem.*, 1946, **18**, 167.
4. H. Green, Industrial Rheology and Rheological Structures, Chapman and Hall (London), 1949.
5. M. Dahlgren, *Trans. Chalmers Univ. Tech.*, 1955, Gothenburg, No. 159, I.
6. R. S. Lenk, *J. Appl. Poly. Sci.*, 1967, **11**, 1033–1042.
7. J. G. Brodnyan, F. H. Gaskins and W. Philippoff, *Trans. Soc. Rheology*, 1957, **1**, 109.
8. W. A. Wright and W. W. Crouse, *ASLE Trans.*, 1965, **8**, 184–190.
9. R. E. Eckert, *J. Appl. Poly. Sci.*, 1963, **7**, 1715–1729.
10. E. B. Bagley, S. H. Story and D. C. West, *Ibid*, 1963, **7**, 1661–1672.
11. K. G. Kendall, *Trans. and J. Plast. Inst.*, June 1963, **31**, 49–59.
12. P. Clegg in 'The Rheology of Elastomers' (Welwyn Garden City Conference 1957), 174 ff. Pergamon Press, P. Mason and N. Wookey, editors.
13. J. P. Tordella, *J. Appl. Poly. Sci.*, 1963, **7**, 215–229.
14. G. W. Scott-Blair, *Rheol. Acta*, March 1965, **4**, 53–55.
15. J. M. McKelvey, Polymer Processing, John Wiley, 1962, 32.
16. M. M. Cross, 'Use of a Flow Equation for the Rheological Characterisation of Pseudoplastic Systems', paper read at the Conference on Advances in Polymer Science and Technology, London, Sept. 1966.
17. M. M. Cross, *J. Colloid Sci.*, 1965, **20**, 417.

Chapter 2

1. R. B. Bird, W. E. Steward and E. W. Lightfoot, Transport Phenomena, John Wiley, 1960.
2. J. M. McKelvey, Polymer Processing, John Wiley, 1962 (Appendix A).
3. *Ibid.*

Chapter 3

1. R. B. Bird, W. E. Steward and E. N. Lightfoot, Transport Phenomena, John Wiley, 1960.

Chapter 4

1. A. A. Miller, *J. Poly. Sci.*, June 1963, **A, 1** (6), 1857–1874.
2. A. K. Doolittle, *J. Appl. Phys.*, **22**, 1031, 1471 (1951); *J. Appl. Phys.*, **23**, 418 (1952).
3. F. Guttmann and L. M. Simmons, *J. Appl. Phys.*, **23**, 977 (1952).
4. M. L. Williams, *J. Appl. Phys.*, **29**, 1395 (1958).
5. M. H. Cohen and D. Turnbull, *J. Chem. Phys.*, **31**, 1164 (1959).
6. R. H. Cole, *Ann. Rev. Phys. Chem.*, **11**, 161 (1960).
7. Dow, Dibert and Fink, *J. Appl. Phys.*, 1939, **10**, 113.
8. P. W. Bridgman, *Revs. Mod. Phys.*, 1946, **18**, 1.
9. J. F. Carley, *Mod. Plastics*, Dec. 1961, with 12 refs.
10. A. K. Van der Vegt and P. P. A. Smit, see Ch. 1, ref (1).
11. R. S. Lenk, see Ch. 1, ref (6).
12. W. C. Schmieder, W. C. Carter, M. Mazat and C. P. Smyth, *J. Am. Chem. Soc.*, 1945, **67**, 959.
13. F. N. Müller, *Kolloid Z.*, Dec. 1953, **134** (2/3), 207.
14. A. B. Bestul and H. V. Belcher, *J. Appl. Phys.*, 1953, **24**, 696.
15. W. Philippoff and F. H. Gaskins, *J. Poly. Sci.*, 1956, **21**, 205.
16. L. R. G. Trelour, The Physics of Rubber Elasticity O.U.P., 1958.

Chapter 5

1. E. H. Merz, R. Kircher and C. W. Hamilton in a paper presented at the Symposium on the Study of Stress/Strain Behaviour, Boston 1960.
2. J. P. Tordella, *J. Appl. Phys.*, 1956, **27**, 404.
3. J. P. Tordella, *SPE Journal*, Feb. 1956, **36**.
4. K. G. Kendall, *Trans. and J. Plast. Inst.*, June 1963, **31** (99), 49–59.
5. J. P. Tordella, *J. Appl. Poly. Sci.*, 1963, **7**, 215–229.
5a. H. J. M. A. Mieras, paper presented at the Autumn Conference of the British Society of Rheology, Shrivenham, Sept. 1967.
6. P. Clegg, see Ch. 1, ref (12).
7. E. B. Bagley, S. H. Storey and D. C. West, *J. Appl. Poly. Sci.*, 1963, **7**, 1661–1672.
8. R. E. Eckert, *Ibid*, 1715–1729.
9. H. P. Schreiber, A. Rudin and Bayley, 1965, *Ibid*, **9**, 887–892.

9a. S. J. Gill and R. Toggenburger, *J. Poly. Sci.*, 1962, 60, F69 ff.

10. H. P. Schreiber and D. C. Storey, *Polymer Letters*, Part B, Sept. 1965, 3 (9), 723–727.

11. W. F. Busse, *Physics Today*, Sept. 1964, 17 (32).

12. J. Galt and B. Maxwell, see Ch. 1, ref. (2).

13. J. J. Benbow and P. Lamb, *S.P.E. Trans.*, 1963, 3, 7.

14. J. P. Tordella, *J. Appl. Phys.*, 1956, 27, 454.

15. F. N. Cogswell and P. Lamb, Trans. & J. Plastics Inst., 35, (120), Dec. 1967, 809–813.

Chapter 6

1. J. F. Carley, *SPE Journal*, Sept. 1963, 19 (9), 977–983. Much of what follows in the text has been culled from this excellent paper.

2. J. F. Carley, *SPE Journal*, Dec. 1963, 19 (12).

3. J. F. Carley, *Mod. Plastics Magazine*, Aug. 1956.

4. J. M. McKelvey, Polymer Processing, Ch. III, sections 5–8 incl., esp. 111–113.

5. Y. Mori and T. K. Matsumoto, *Rheol. Acta*, 1958, 1, 240

6. R. M. Griffiths, *Ind. Eng. Chem. (Funds)*, 1962, 1, 180.

7. R. G. Gaskell, *J. Appl. Mechanics*, 1950, 17, 334.

8. R. B. Gee and J. B. Lyon, *Ind. Eng. Chem.*, 1957, 49, 956.

9. J. Pearson, *Trans. and J. Plast. Inst.*, 1962, 30, 230.

10. J. Pearson, *Ibid*, 1963, 31, 125.

11. J. Pearson, *Ibid*, 1964, 32, 239.

Chapter 7

1. A. V. Tobolsky, Properties and Structure of Polymers, 31–36. Wiley, 1960.

2. Turner Alfrey, Mechanical Behaviour of High Polymers, High Polymer Series Vol. VI, Interscience 1948.

3. Wiechert, *Ann. Physik*, 1893, 50, 335, 546.

4. Smekal, *Z. phys. Chem.*, 1939, B44, 286.

5. T. Alfrey and P. Doty, *J. Appl. Phys.*, 1945, 16, 700.

6. S. Turner, *Trans. and J. Plastics Inst.*, June 1966, 34 (11) 127–135.

7. A. V. Tobolsky and H. Eyring, *J. Chem. Phys.*, 1943, 11, 125 (and others subsequently).

8. S. Turner, *Brit. Plastics*, June 1964, 37 (6), 322–324.

9. C. M. R. Dunn, W. H. Mills and S. Turner, *Brit. Plastics*, July 1964.

10. H. J. Scherr and W. E. Palm, *J. Appl. Poly. Sci.*, 1963, 7, 1273–1279.

11. R. B. Stambough, *Ind. Eng. Chem.*, 1952, 44, 1590.

12. G. Ajroldi, C. Garbuglio and G. Pezzin, *J. Poly. Sci.*, 1967, A–Z, 5, 289–300.

13. F. W. Schremp, J. D. Ferry and W. W. Evans, *J. Appl. Phys.*, 1951, **22**, 711.
14. J. D. Ferry and M. L. Williams, *J. Colloid Sci.*, 1952, **7**, 347.
15. J. A. Faucher, *J. Appl. Phys.*, 1961, **32**, 2336.
16. H. Eyring and T. Ree, *J. Appl. Phys.*, 1955, **26**, 793.

Chapter 8

1. P. Vincent in 'Physics of Plastics', Ch. I, Iliffe 1965, (Prof. Ritchie, editor).
2. P. Vincent, *Polymer*, 1959, **1**, 7.
3. After P. Vincent, *Plastics*, Oct. 1961, **26**, 121.
4. P. Vincent, *Plastics*, Nov. 1961, **26**, 141.
5. N. Goldenberg, M. Arcan and E. Nicolau in Proc. of the International Symposium on Plastics Testing and Standardisation, Philadelphia 1958.
6. P. Vincent, *Plastics*, Jan. 1962, **27**, 115.
7. E. R. Parker, Brittle Behaviour of Engineering Structures, John Wiley, 1957.
8. S. Turner and *J. Plast. Inst.*, Aug. 1965, **33**, 106.

Chapter 9

1. C. J. F. Böttcher, Theory of Electric Polarisation, Ch. X, Elsevier 1952. This contains a detailed derivation of the Debye Equations.
2. H. Fröhlich, Theory of Dielectrics, Clarendon Press (Oxford), 1949.
3. W. Reddish, *Trans. Faraday Soc.*, June 1950, **46** (6), 330.
4. K. Deutsch, E. A. W. Hoff and W. Reddish, *J. Poly, Sci.*, 1951, **13** (72), 565–582.
5. F. Würstlin and H. Thurn, Die Physik der Hochpolymeren, Vol. VI. H. A. Stewart (editor), Springer 1956.
6. C. Mussa, *Trans. and J. Plast. Inst.*, Dec. 1963, **31** (96), 146–147.
7. G. C. Karas, Principles of Dynamic Testing, Brit. Plastics, Feb. 1964, 59–62.
8. K. H. Illers and E. Jenckel, *Kolloid Z.*, Oct. 1958, **160** (2), 98.
9. T. Raphael and C. D. Armeniades, *SPE Trans.*, April 1964, **4**, 2.
10. I. Cheetham, *Trans. Proc. Rubber Ind.*, 1965, **40** (4), T156–179.